# Multiplier-Cum-Divider Circuits

# Multiplier-Cum-Divider Circuits

Principles, Design, and Applications

K. C. Selvam

CRC Press is an imprint of the
Taylor & Francis Group, an **informa** business

First edition published 2021 by
CRC Press
6000 Broken Sound Parkway NW, Suite 300, Boca Raton, FL 33487-2742

and by CRC Press
2 Park Square, Milton Park, Abingdon, Oxon OX14 4RN

© 2022 K. C. Selvam

CRC Press is an imprint of Taylor & Francis Group, LLC

Reasonable efforts have been made to publish reliable data and information, but the author and publisher cannot as sume responsibility for the validity of all materials or the consequences of their use. The authors and publishers have attempted to trace the copyright holders of all material reproduced in this publication and apologize to copyright holders if permission to publish in this form has not been obtained. If any copyright material has not been acknowledged please write and let us know so we may rectify in any future reprint.

Except as permitted under U.S. Copyright Law, no part of this book may be reprinted, reproduced, transmitted, or utilized in any form by any electronic, mechanical, or other means, now known or hereafter invented, including photocopying, microfilming, and recording, or in any information storage or retrieval system, without written permission from the publishers.

For permission to photocopy or use material electronically from this work, access www.copyright.com or contact the Copyright Clearance Center, Inc. (CCC), 222 Rosewood Drive, Danvers, MA 01923, 978-750-8400. For works that are not available on CCC please contact mpkbookspermissions@tandf.co.uk

*Trademark notice*: Product or corporate names may be trademarks or registered trademarks and are used only for identification and explanation without intent to infringe.

ISBN: 978-0-367-75446-4 (hbk)
ISBN: 978-1-003-16851-5 (ebk)
ISBN: 978-0-367-76779-2 (pbk)

Typeset in Times
by Newgen Publishing UK

*Dedicated to my loving wife*
*S. Latha*

# Contents

Preface ............................................................................................................. xi
Author biography ........................................................................................ xiii
List of Abbreviations .................................................................................... xv

**Chapter 1**  Introduction ............................................................................ 1

              1.1  Characteristics ................................................................. 3
              1.2  Specifications .................................................................. 3
              1.3  Errors ............................................................................... 3
              1.4  Individual Error ............................................................... 4
              1.5  Offset Nulling .................................................................. 5
              1.6  MCD Types ..................................................................... 5

**Chapter 2**  Basic Components ................................................................. 7

              2.1  Inverting Amplifier .......................................................... 7
              2.2  Non-inverting Amplifiers ................................................ 8
              2.3  Integrator ....................................................................... 10
              2.4  Analog Switches ........................................................... 11
              2.5  Analog Multiplexers ...................................................... 14
              2.6  Astable Multivibrator .................................................... 15

**Chapter 3**  Non-linear Circuits .............................................................. 19

              3.1  Voltage Comparators .................................................... 19
              3.2  Schmitt Triggers ............................................................ 24
              3.3  Half-Wave Rectifiers .................................................... 25
              3.4  Full-Wave Rectifiers ..................................................... 26
              3.5  Peak Detectors .............................................................. 29
              3.6  Sample and Hold Circuits ............................................. 30
              3.7  Log Amplifiers .............................................................. 30
              3.8  Anti-log Amplifiers ....................................................... 32

**Chapter 4**  Conventional MCDs ............................................................ 37

              4.1  Log–Anti-log MCDs – Type I ...................................... 37
              4.2  Log–Anti-log MCDs – Type II ..................................... 40
              4.3  MCD Using FETs .......................................................... 41
              4.4  MCD Using MOSFETs .................................................. 43
              4.5  MCD Using Two Analog Multipliers ............................ 45
              4.6  MCD Using Two Analog Dividers ................................ 46

|  |  |  |
|---|---|---|
| | 4.7 | MCD Using Two Analog Dividers in Cascade ........................ 46 |
| | 4.8 | MCD Using a Divider and a Multiplier in Series .................... 47 |

**Chapter 5** Sawtooth Wave-Referenced Time-Division MCD Using Multiplexers.................................................................................... 49

- 5.1 Sawtooth Wave Generators ........................................................ 49
- 5.2 Double Multiplexing–Averaging Time-Division MCD .......... 52
- 5.3 Time-Division Single-Slope Peak-Detecting MCD................. 55
- 5.4 Time-Division Multiply–Divide MCD .................................... 58
- 5.5 Time-Division Divide–Multiply MCD .................................... 61

**Chapter 6** Triangular Wave-Referenced MCD with Multiplexers ...................... 65

- 6.1 Triangular Wave Generators...................................................... 65
- 6.2 Time-Division MCD ................................................................. 68
- 6.3 Time-Division Divide–Multiply MCD .................................... 70
- 6.4 Time-Division Multiply–Divide MCD .................................... 73

**Chapter 7** Peak-Responding MCDs with Multiplexers....................................... 77

- 7.1 Double Single-Slope Peak-Responding MCD......................... 77
- 7.2 Double Dual-Slope Peak-Responding MCD Using Feedback Comparator .............................................................. 81
- 7.3 Double Dual-Slope Peak-Responding MCD with Flip Flop...................................................................................... 85
- 7.4 Pulse-Width Integrated Peak-Responding MCD ..................... 88
- 7.5 MCD Using Voltage Tunable Astable Multivibrator................ 91

**Chapter 8** Sawtooth Wave-Referenced MCDs Using Analog Switches ............. 95

- 8.1 Sawtooth Wave Generators ....................................................... 95
- 8.2 Double Switching and Averaging Time-Division MCD .......... 97
- 8.3 Time-Division Single-Slope Peak-Detecting MCD............... 101
- 8.4 Time-Division Multiply–Divide MCD .................................. 104
- 8.5 Time-Division Divide–Multiply MCD .................................. 109

**Chapter 9** Triangular Wave-Referenced MCD with Analog Switches.............. 115

- 9.1 Time-Division MCD ............................................................... 115
- 9.2 Divide–Multiply Time-Division MCD .................................. 119
- 9.3 Multiply–Divide Time-Division MCD .................................. 123

**Chapter 10** Peak-Responding MCDs with Analog Switches ............................. 129

- 10.1 Double Single-Slope Peak-Responding MCDs ..................... 129
- 10.2 Double Dual-Slope MCD Using Feedback Comparator ........ 133
- 10.3 Double Dual-Slope MCD with Flip Flop............................... 140

Contents ix

        10.4  Pulse-Width Integrated Peak-Responding MCD ................... 145
        10.5  MCD Using Voltage-Tunable Astable Multivibrator ............ 149

**Chapter 11**  Time-Division MCD without Reference .......................................... 155

        11.1  Time-Division MCD Type I – Multiplexing ......................... 155
        11.2  Time-Division MCD Type II – Multiplexing........................ 158
        11.3  Time-Division MCD Type I – Switching............................... 161
        11.4  Time-Division MCD Type II – Switching ............................ 166

**Chapter 12**  Pulse Position-Responding MCDs .................................................... 173

        12.1  Pulse Position Peak-Detecting MCDs – Multiplexing........... 173
        12.2  Pulse Position Peak-Detecting MCDs – Switching .............. 177
        12.3  Pulse Position Peak-Sampling MCD – Multiplexing ............ 180
        12.4  Pulse Position Peak-Sampling MCD – Switching................ 183

**Chapter 13**  Applications of MCDs ..................................................................... 187

        13.1    Balanced Modulator............................................................. 187
        13.2    Amplitude Modulator........................................................... 188
        13.3    Frequency Doubler............................................................... 189
        13.4    Phase Angle Detector........................................................... 190
        13.5    RMS Detector ...................................................................... 191
        13.6    Rectifier................................................................................ 191
        13.7    Inductance Measurement by Phase Angle Response............ 192
        13.8    Capacitance Measurement by Phase Angle Response.......... 193
        13.9    Automatic Gain Control Circuit – Type I ............................. 195
        13.10  Automatic Gain Control Circuit – Type II............................ 195

**Chapter 14**  Circuit Simulation ........................................................................... 197

        14.1  Simulation of Time-Division Multiply–Divide MCD............ 198
        14.2  Simulation of Time-Division Divide–Multiply MCD............ 200
        14.3  Simulation of Time-Division MCD – Type I –
              Switching ............................................................................. 200
        14.4  Simulation of Time-Division MCD – Type II –
              Switching ............................................................................. 202

Appendix.......................................................................................................... 207
Bibliography ................................................................................................... 211
Index ................................................................................................................ 215

# Preface

After publishing one book on multipliers and one book on dividers, I came up with the concept for a multiplier-cum-divider (MCD). This MCD will work as a multiplier as well as a divider. In the year 2000, I published (i) first, a MCD circuit in *IETE Journal of Education* (Vol. 41, Nos 1 and 2, Jan–June 2000) (ii) second, an MCD circuit in *Electronics Letters Journal* (Vol. 49, No. 23, Nov. 2013) and (iii) third, an MCD circuit in *IETE Journal of Research* (28 Feb 2018). In 2019, I designed 127 MCD circuits, and decided to write a book, and this is the result.

I am highly indebted to:

(i) Mentor Prof. Dr. V.G.K. Murti who taught me how to get the average value of a periodic waveform
(ii) Philosopher Prof. Dr. P. Sankaran who introduced me to the IIT Madras
(iii) Teacher Prof. Dr. K. Radha Krishna Rao who taught me operational amplifiers
(iv) Director Prof. Dr. Bhaskar Ramamurti who motivated me to do this work
(v) Gurunather Prof. Dr. V. Jagadeesh Kumar who guided me in the proper way of the scientific world
(vi) Trainer Dr. M. Kumaravel who trained me to do experiments with operational amplifiers
(vii) Prof. Dr. Enakshi Bhattacharya who encouraged me to get useful results
(viii) Leader Prof. Dr. Devendra Jalihal who kept me in a happy and peaceful official atmosphere
(ix) Supervisor Prof. Dr. David Koil Pillai who supervised and monitored all my research work.

I am indebted to Dr. S. Sathiyanathan, Dr. R. Rajkumar and Dr. V. Vasantha Jayaram who gave me medical treatment during the years 1996–2010. I am also indebted to Dr. Shiva Prakash, psychiatrist, and Dr. Saraswathi, general physician, who continue the medical treatment.

I am indebted to Prof. Dr. T.S. Rathore, former professor of IIT Bombay who reviewed all my papers and Prof. Dr. Raj Senani, former editor of *IETE Journal of Education*, who published most of my papers.

I thank my father, Mr. Venkatappa Chinthambi Naidu, mother, Mrs. C. Suseela, wife, S. Latha, elder son, S. Devakumar, and younger son, S. Jagadeesh Kumar, for keeping me in a happy and peaceful residential atmosphere.

I thank my friends Prof. Dr. R. Sarathi, Prof. Dr. K. Sridharan, Dr. Bharath Bhikkaji, Dr. Boby George, and Dr. Anirudhan for encouraging me throughout my research work.

I thank Dr. Gagandeep Singh, Senior Editor, Taylor and Francis, CRC Press, who has shown a keen interest in publishing all my concepts on function circuits. His hard work in bringing this book is commendable.

I thank my apprentice trainee classmates and friends Mrs. T. Padmavathy, Mr. V. Selvaraju, Dr. C.R. Jeevandos and Mrs. P.V. Suguna, with whom I started my career at the Central Electronics Centre of IIT Madras in 1987.

**K.C. Selvam**

# Author biography

**K.C. Selvam** is currently working as a technical officer, department of electrical engineering, Indian Institute of Technology Madras, India. He has conducted research and development work for the last 30 years and has published more than 34 research papers in various national and international journals. He published a book entitled *Design of Analogue Multipliers with Operational Amplifiers* with CRC Press, Taylor & Francis. He received a best paper award by IETE in 1996. He received the student's journal award of IETE in 2017. He developed an interest in design and development of function circuits to find their applications in modern measurements and instrumentation systems.

# List of Abbreviations

$V_1$    First input voltage
$V_2$    Second input voltage
$V_3$    Third input voltage
$V_O$    Output voltage
$V_R$    Reference voltage / Peak value of first sawtooth waveform
$V_T$    Peak value of first triangular waveform
$V_P$    Peak value of second triangular wave / sawtooth wave
$V_C$    Comparator 1 output voltage in the first sawtooth / triangular wave generator
$V_M$    Comparator 2 output voltage by comparing sawtooth / triangular waves with one input voltage
$V_N$    Low pass filter input signal
$V_{S1}$    First generated sawtooth wave
$V_{S2}$    Second generated sawtooth wave
$V_{T1}$    First generated triangular wave
$V_{T2}$    Second generated triangular wave
$V_S$    Sampling pulse
$V_1'$    A voltage that is slightly less than $V_1$
$V_2'$    A voltage that is slightly less than $V_2$

# 1 Introduction

The world of electronics is all about electrical circuits, electronic components, and interconnected technologies. All these elements can be primarily categorized as digital, analog, or a combination of both. However, here we will be focusing on the basics of the analog category in detail.

Analog electronics is a branch of electronics that deals with a continuously variable signal. It's widely used in radio and audio equipment along with other applications where signals are derived from analog sensors before being converted into digital signals for subsequent storage and processing. Although digital circuits are considered as a dominant part of today's technological world, some of the most fundamental components in a digital system are actually analog in nature.

Analog means continuous and real. The world we live in is analog in nature, implying that it's full of infinite possibilities. The number of smells we can sense, the number of tones we can hear, or the number of colors we can paint with; everything is infinite. The people working in the field of analog electronics are basically dealing with analog devices and circuits.

Analog signals: Before proceeding with analog signals, let's understand the simple meaning of a signal. In electrical engineering, signals are basically time-varying quantities (usually voltage or current). So when we talk about the signal it means we are talking about a voltage that's changing over time.

Signals are passed among devices in order to obtain or send information in the form of audio, video, or encoded data. The transmission takes place through wires or via air through radio frequency waves. For instance, audio signals are transferred from the computer's audio card to the speakers, while data signals between a tablet and a Wi-Fi router pass through the air.

Analog signals use attributes of the medium to convey the signal's information. For example, an aneroid barometer makes use of the angular position of a needle to convey the changes in atmospheric pressure. The signals take any value from a given range and each signal value denotes different information. Each level of the signal signifies a different level of the phenomenon and any change in the signal is meaningful.

It is easy to determine a signal input as analog or digital. An analog signal is smooth and continuous. A digital signal is in the form of stepping squares.

Analog circuits can be defined as a complex combination of op-amps, resistors, caps, and other basic electronic components. These circuits can be as simple as a combination of two resistors to make a voltage divider or elegantly built with many components. Such circuits can attenuate, amplify, isolate, modify, distort the signal, or even convert the original one into a digital signal.

These circuits are difficult to design and need a lot of precision compared to digital circuits. Modern circuits are rarely found to be completely analog as these days analog circuitry may use digital or microprocessor techniques to improve performance. Such circuits are called mixed signals.

There are two types of analog circuits, namely passive and active, where the former don't consume any electrical power while the latter do.

The most basic difference between analog and digital electronics is that in the former technology translates the information into electric pulses of varying amplitude, while the latter translates information into a binary format of 0 and 1, where each bit represents two distinct amplitudes.

Operational amplifiers play an important role in the field of analog electronics. This book describes how op-amps can be used to develop multiplier-cum-divider (MCD) circuits.

An MCD is an analog function circuit that accepts three input voltages $V_1, V_2, V_3$ and produces an output voltage $V_O = \dfrac{V_2 V_3}{V_1}$. Its symbol is shown in Fig. 1.1.

The MCD is very useful in instrumentation and communication applications. It can be used in many applications, such as analog signal processing (modulators, demodulators), automatic gain control, frequency translation, controlled resistors, analog computation, fuzzy-control, instrumentation (wattmeters, impedance meters, flow meters, RMS-DC converters), etc.

Analog MCDs are used for frequency conversion and are critical components in modern radio frequency (RF) systems. RF systems must process analog signals with a wide dynamic range at high frequencies. A mixer converts RF power at one frequency into power at another frequency to make signal processing easier and also inexpensive. A fundamental reason for frequency conversion is to allow amplification of the received signal at a frequency other than the RF, or the audio, frequency.

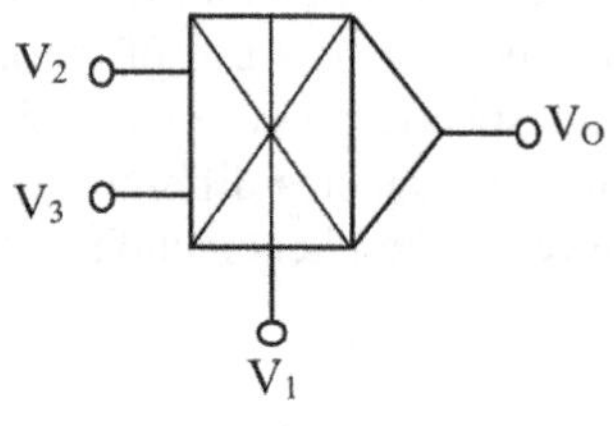

**FIGURE 1.1** MCD symbol

# Introduction

## 1.1 CHARACTERISTICS

An ideal MCD will have (i) infinite input impedance and (ii) zero output impedance. But in practice no MCD meets these characteristics 100%. The input voltages have a finite differential and common mode voltage range with finite impedance. The output has a finite current output capacity and finite output impedance.

## 1.2 SPECIFICATIONS

The following are specifications of an MCD

(i) Rated output (minimum voltage at rated current)
(ii) Output impedance
(iii) Maximum input voltage
   (a) Maximum voltage for rated specification
   (b) Maximum voltage for no damage to device.
(iv) Input impedance
(v) 3 dB band width
(vi) Output slew rate
(vii) Output settling time.

## 1.3 ERRORS

When MCD is configured to work as multiplier, the actual output of this multiplier can be expressed as

$$Vo = (k_m + \Delta k_m)(V_1 + V_{1IO})(V_2 + V_{2IO}) + V_{oo} + V_1 f + V_2 f + V_{n(1,2)} \quad\quad (1.0)$$

Where

$\Delta k_m$ = Scale factor error
$V_{1IO}$ = Input offset voltage of $V_1$ input
$V_{2IO}$ = Input offset voltage of $V_2$ input
$V_{oo}$ = Output offset voltage
$V_1 f$ = Non-linear feedthrough from $V_1$ input to $V_o$ output
$V_2 f$ = Non-linear feedthrough from $V_2$ input to $V_o$ output
$V_n(1,2)$ = Non-linearity in gain response.

Rearrange the above expression (1.0) given

$$V_o = K_m v_{1+} v_2 + \Delta_m(v_{1+} v_2) + [(K_m + \Delta k_m)V1V2IO+V1f]$$
$$+ [(k_m + \Delta k_m)(V_2 V_{1IO}) + (V_{oo} + V_n)_{(1,2)}$$

$K_m v_{1+} v_2 \rightarrow$ ideal output
$\Delta_m(v_{1+} v_2) \rightarrow$ gain error
$(K_m + \Delta k_m)V1V2IO+V1f \rightarrow$ total $V_1$ – input feedthrough

$(k_m + \Delta k_m)(V_2 V_{1Io}) \rightarrow$ total $V_2$ input feed though
$V_{oo} \rightarrow$ output offset
$(V_{oo} + V_n)_{(1,2)} \rightarrow$ non-linear output

The multiplier has four external adjustments

(i) The $V_1$ input offset null (for $V_{1Io}$)
(ii) The $V_2$ input offset null (for $V_{2Io}$)
(iii) The output offset null (for $V_{oo}$)
(iv) A gain adjustment (for $V_n$ (1,2)).

## 1.4  INDIVIDUAL ERROR

When any of the numerator inputs of an MCD is zero, and the denominator input voltage is constant, the output is supposed to be zero. But in practical MCDs, the following errors will occur: (i) output offset voltage, (ii) $V_2$ input feedthrough, which is a small error feeding to the output from the $V_2$ input when the $V_3$ input is zero, and (iii) $V_3$ input feedthrough, which is a small error feeding to the output from the $V_3$ input when the $V_2$ input is zero. The output offset voltage is the output voltage when both input voltages $V_2$ and $V_3$ are zero. This can be nulled with suitable offset adjustment presets in the op-amps used for multiplication. The $V_2$ input feedthrough has (i) $V_2$ input signal being multiplied by a finite $V_3$ input. This can be nulled in the $V_3$ input and (ii) varies non-linearly with the $V_2$ input signal and cannot be nulled. The feedthrough error can be minimized by nulling input offset voltages of both input voltages. The feedthrough error will increase due to input offset temperature drifts and hence the multiplier should have self-nulling facilities. The feedthrough error also depends on frequency and the error will increase if frequency is increased.

There may be a difference between $V_2$ input though and $V_3$ input through errors. The feedthrough can be expressed as:

$$Feed\ through = \frac{Peak\ to\ peak\ output\ voltage}{Either\ V_2\ or\ V_3\ input\ at\ zero}$$

The typical valve of feedthrough is 50 mV (0.5%) or less.

Gain error is the output error when all input voltages $V_2, V_3$ are at maximum values (let us say $V_2 = V_3 = 10\ V_{max}$) for both output offset and input offsets and nulled.

A non-linearity error is the maximum deviation of input and output responses from ideal characteristics. Typical commercial multipliers have a non-linearity error between 0.01% and 0.5%.

There are two dynamic errors: (i) distortion, which increases at higher frequencies, and (ii) bandwidth, which depend on the DC signal levels.

Introduction

## 1.5 OFFSET NULLING

The following steps are to be followed for offset nulling in an MCD.
Let us first keep the $V_1$ voltage at a constant value.

(i) Connect the $V_2$ input to the GND (ground). Apply a low-frequency sine wave of maximum amplitude to the $V_3$ input (e.g. 10 Vmax at 50 Hz). Null the $V_2$ offset until the output equals zero.
(ii) Connect the $V_3$ input to the GND. Apply a low frequency sine wave of maximum amplitude to the $V_2$ input. Null the $V_3$ offset until the output equals zero.
(iii) Connect both the $V_2$ input and the $V_3$ input to GND. Null the output offset for zero output voltage.

Next, we will keep the $V_3$ at a constant value.
Connect the $V_1$ input to the value of the $V_3$. Connect the $V_2$ input to the GND. Null the $V_2$ offset until the output becomes zero.

## 1.6 MCD TYPES

MCD circuits are broadly divided into (i) time-division MCDs and (ii) peak-responding MCDs. The peak-responding MCDs are further classified into (i) peak-detecting MCDs and (ii) peak-sampling MCDs. Based on these two types, the following MCDs can be obtained.

(i) Sawtooth wave-referenced time-division MCD with multiplexers
(ii) Triangular wave-referenced MCD with multiplexers
(iii) Peak-responding MCD with multiplexers
(iv) Sawtooth wave-referenced MCD with analog switches
(v) Triangular wave-referenced MCD with analog switches
(vi) Peak-responding MCD with analog switches
(vii) Time-division MCD with no reference
(viii) Pulse position responding MCDs

All these types of MCDs are discussed in this book.

# 2 Basic Components

An inverting amplifier, non-inverting amplifier, integrator, comparator, low pass filter, analog switch, analog multiplexers, astable multivibrator, peak detector and sample and hold circuits are the basic components of a multiplier cum divider (MCD). All MCD circuits are developed using any of the above components. These components are discussed in this chapter briefly.

## 2.1 INVERTING AMPLIFIER

Fig. 2.1 shows an inverting amplifier using an op-amp. Since the non-inverting terminal (+) of an op-amp is grounded through the resistor $R_p$, the voltage $V_A$ at the non-inverting terminal (+) of the op-amp is zero volts. The op-amp is in a negative closed loop feedback and hence its non-inverting terminal voltage will be equal to its inverting terminal voltage, i.e.

$$V_A = V_B = 0$$

The current through resistor R1 will be

$$I = \frac{V_I - V_B}{R_1} = \frac{V_I}{R_1} \tag{2.1}$$

Since the op-amp has infinite input impedance, the current I does not enter the op-amp and flows through the resistor $R_2$. The voltage across resistor $R_2$ will be

$$V_{R2} = IR_2 = \frac{V_I}{R_1} R_2 \tag{2.2}$$

The negative feedback forces the op-amp to produce an output voltage that maintains a virtual ground at the op-amp inverting input. The output voltage is given as

$$V_O = V_B - V_{R2} = -V_{R2}$$

$$V_O = -\left(\frac{R_2}{R_1}\right) V_I \tag{2.3}$$

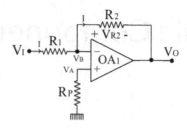

**FIGURE 2.1** Inverting amplifier

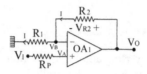

**FIGURE 2.2** Non-inverting amplifier

If $R_1 = R_2$ then

$$V_O = -V_I \tag{2.4}$$

## 2.2 NON-INVERTING AMPLIFIERS

Fig. 2.2 shows a non-inverting amplifier – type I. The op-amp is at a negative closed loop feedback and its inverting terminal (−) voltage will be equal to its non-inverting terminal (+) voltage. i.e.

$$V_A = V_B = V_I$$

The current through resistor $R_1$ will be

$$I = \frac{V_B}{R_1} = \frac{V_I}{R_1} \tag{2.5}$$

The current I comes from resistor $R_2$ and it does not enter the op-amp as the op-amp has infinite input impedance. The voltage across the feedback resistor $R_2$ will be

$$V_{R2} = IR_2 = \frac{V_I}{R_1} R_2 \tag{2.6}$$

The output voltage is given as

$$V_O = V_B + V_{R2}$$

# Basic Components

$$V_O = V_I + \left(\frac{R_2}{R_1}V_I\right)$$

$$V_O = V_I\left(1 + \frac{R_2}{R_1}\right) \qquad (2.7)$$

If $R_1 = R_2$, then

$$V_O = 2V_I \qquad (2.8)$$

Fig. 2.3 shows a unity gain non-inverting amplifier – type II. Let us analyze the circuit using the superposition principle.

First, the non-inverting terminal (+) is grounded through the resistor $R_P$ and the input voltage $V_I$ is given to the inverting terminal through the resistor $R_1$. As discussed in section 2.1, the circuit will work as an inverting amplifier and if $R_1 = R_2$, the output voltage will be

$$V_{O1} = -V_I \qquad (2.9)$$

Next, the inverting terminal (−) is grounded through resistor R1 and the input voltage is given to the non-inverting terminal (+) through resistor RP. As discussed at the start of section 2.2, the circuit will work as a non-inverting amplifier and if $R_1 = R_2$, the output voltage will be

$$V_{O2} = 2V_I \qquad (2.10)$$

By superposition principle the actual output voltage will be addition of equation (2.9) and (2.10)

$$V_O = V_{O1} + V_{O2} \qquad (2.11)$$

$$V_O = V_I \qquad (2.12)$$

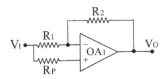

**FIGURE 2.3** Unity gain non-inverting amplifier

## 2.3 INTEGRATOR

Fig. 2.4(a) shows an integrator using an op-amp. Since the non-inverting terminal (+) is at ground potential, the inverting terminal (−) will also be at ground potential by virtual ground. The current through the resistor R will be

$$I = \frac{V_I - 0}{R} = \frac{V_I}{R} \tag{2.13}$$

Due to the op-amp's high input impedance, the current I will not enter in to op-amp and flows through the capacitor C. The voltage across capacitor C will be

$$V_C = \frac{q}{C} \tag{2.14}$$

In the above equation (2.14), 'q' is charged to exist on the plates of the capacitor. The relation between current (i) and charge (q) is given as

$$i = \frac{dq}{dt}$$

$$q = \int I \, dt \tag{2.15}$$

Equation (2.15) in (2.14) gives

$$V_C = \frac{\int I \, dt}{C} \tag{2.16}$$

The negative feedback forces the op-amp to produce an output voltage that maintains a virtual ground at the op-amp inverting input. The output voltage is given as

$$V_O = 0 - V_C = -V_C \tag{2.17}$$

equation (2.16) in (2.17) gives

$$V_O = -\frac{1}{C} \int I \, dt \tag{2.18}$$

Equation (2.13) in (2.18) gives

$$V_O = -\frac{1}{RC} \int V_I \, dt = -\frac{V_I}{RC} t \tag{2.19}$$

Fig. 2.4(b) shows circuit of a practical integrator. If a square wave is given as the input to the integrator, then a triangular wave will be the output of the integrator.

Basic Components

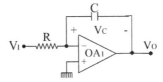

**FIGURE 2.4(A)** Integrator

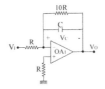

**FIGURE 2.4(B)** Practical integrator

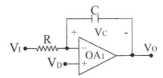

**FIGURE 2.4(C)** Differential integrator

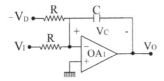

**FIGURE 2.4(D)** Equivalent circuit of Fig. 2.4(c)

Fig. 2.4(c) shows a differential integrator. Its output is given as

$$V_O = \frac{1}{RC}\int (V_D - V_I)dt = \frac{(V_D - V_I)}{RC}t \qquad (2.20)$$

Fig 2.4(d) shows an equivalent circuit of the differential integrator shown in Fig. 2.4(c).

## 2.4 ANALOG SWITCHES

The symbol of an analog switch is shown in Fig. 2.5. It has three terminals, CON, IN/OUT, and OUT/IN. If the control (CON) pin is LOW, the switch $S_1$ is opened so that the IN/OUT and OUT/IN terminals are disconnected. If the control (CON) pin is HIGH, the switch $S_1$ is closed so that the IN/OUT and OUT/IN terminals are connected together.

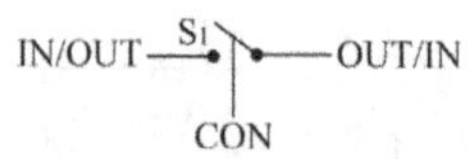

**FIGURE 2.5**  Switch symbol

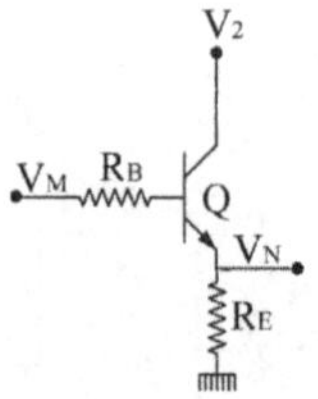

**FIGURE 2.6(A)**  Transistor series switch

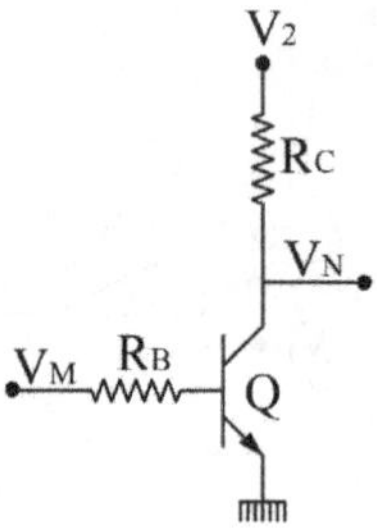

**FIGURE 2.6(B)**  Transistor shunt switch

Analog switches are available in an IC PACKAGE of CMOS CD4066 IC. The pin details of this CD4066 IC are given in the appendix.

Fig. 2.6(a) shows a transistor series switch. If the control input $V_M$ is LOW, the transistor Q is OFF, and the collector voltage will not exist on the emitter voltage

$$V_N \sim 0$$

If the control input $V_M$ is HIGH, the transistor Q is ON, and the collector voltage will exist on the emitter terminal.

$$V_N \sim V_2$$

Fig. 2.6(b) shows a transistor shunt switch. If the control input $V_M$ is LOW, the transistor Q is OFF, $V_N \sim V_2$ ($R_C$ value is very low). If the control input $V_M$ is HIGH, the transistor Q is ON, $V_N \sim 0$.

Fig. 2.7(a) shows the Field Effect Transistor (FET) as a series switch. If the control input is HIGH(+Vcc), zero volts will exist on the gate terminal, and FET is ON and acts as a closed switch. OUT ~ IN. If the control input CON is LOW(−Vcc), a negative voltage will exist on the gate terminal, and FET is OFF and acts as an open switch. OUT ~ 0.

# Basic Components

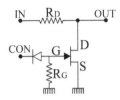

**FIGURE 2.7(A)**   FET series switch

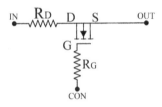

**FIGURE 2.7(B)**   FET shunt switch

**FIGURE 2.8(A)**   MOSFET series switch

Fig. 2.7(b) shows a FET shunt switching circuit. If the control input CON is HIGH(+Vcc), zero volts will exist on gate, and FET is ON and zero volts will be output. OUT ~ 0. If the control input CON is LOW(−Vcc), a negative voltage will exist on the gate terminal, and the FET operated on a cut-off region acts as an open circuit. The output will be OUT ~ IN.

Fig. 2.8(a) shows a MOSFET (metal–oxide–semiconductor field-effect transistor) series switch. If the control input is HIGH(+$V_{DD}$), the channel resistance becomes very small and allows the maximum drain current to flow. This is the saturation mode and the MOSFET is completely ON and acts as a closed circuit OUT ~ IN. If the control input is LOW($V_{SS}$), the channel resistance become HIGH and no current flows from the drain. This is a cut-off region and the MOSFET is completely OFF and acts as an open switch. OUT ~ 0.

Fig. 2.8(b) shows a MOSFET shunt switch. If the control input is HIGH(+$V_{DD}$), the channel resistance becomes very small and allows the maximum drain current to flow. This is the saturation mode and the MOSFET is completely ON and acts as a closed circuit. OUT ~ 0. If the control input is LOW(VSS), the channel resistance become HIGH and no current flows from the drain. This is a cut-off region and the MOSFET is completely OFF and acts as an open switch. OUT ~ IN.

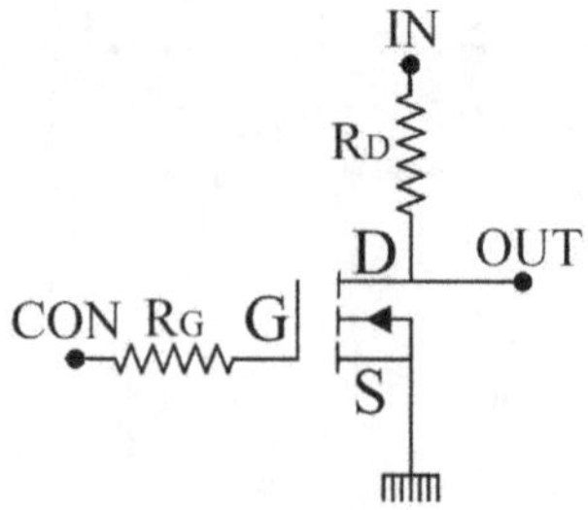

**FIGURE 2.8(B)** MOSFET shunt switch

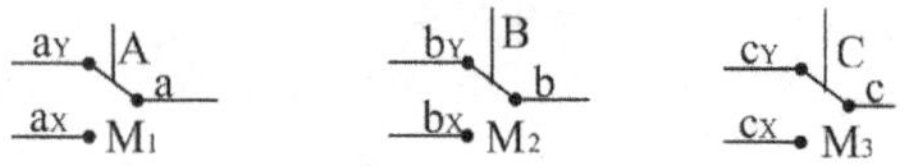

**FIGURE 2.9** Triple two-to-one multiplexers

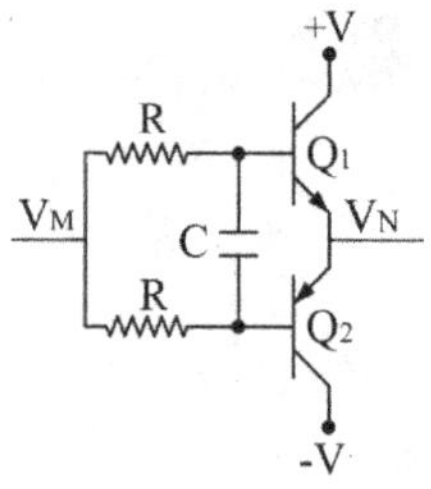

**FIGURE 2.10** Transistor multiplexer

## 2.5 ANALOG MULTIPLEXERS

Fig. 2.9 shows the symbol of an analog triple 2 to 1 multiplexer. Each multiplexer has four terminals. In the case of multiplexer $M_1$, it has 'ay,' 'ax,' 'a' and 'A' terminals. In the case of multiplexer $M_2$, it has 'by,' 'bx,' 'b' and 'B' terminals. In the case of multiplexer $M_3$, it has 'cy,' 'cx,' 'c' and 'C' terminals.

In multiplexer $M_1$, if the pin 'A' is HIGH, then 'ay' is connected to 'a' and if the pin 'A' is LOW, then 'ax' is connected to 'a.' In multiplexer $M_2$, if the pin 'B' is HIGH, then 'by' is connected to 'b' and if the pin 'B' is LOW, then 'bx' is connected to 'b.' In multiplexer $M_3$, if the pin 'C' is HIGH, then 'cy' is connected to 'c' and if the pin 'C' is LOW, then 'cx' is connected to 'c.'

All the three multiplexers $M_1, M_2$ and $M_3$ are available in one IC PACKAGE of CMOS CD4053 IC. The pin details of this CD4053 IC are given in the appendix.

Figure 2.10 shows analog multiplexer using transistors. If the control input $V_M$ is HIGH $(+V_{CC})$, the transistor $Q_1$ is ON, $Q_2$ is OFF and $(+V)$ will appear at $V_N$. If the control input CON is LOW $(-V_{CC})$, the transistor $Q_1$ is OFF, $Q_2$ is ON and $(-V)$ will appear at $V_N$.

Basic Components 15

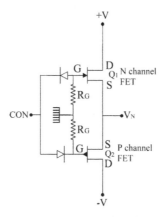

**FIGURE 2.11** FET multiplexer

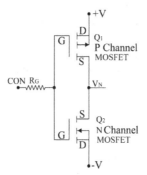

**FIGURE 2.12** MOSFET multiplexer

Figure 2.11 shows an analog multiplexer using FETs. If the control input CON is HIGH (+$V_{CC}$), FETs $Q_1$ is ON, $Q_2$ is OFF and (+V) will appear at $V_N$. If the control input CON is LOW (−$V_{CC}$), FETs $Q_1$ is OFF, $Q_2$ is ON and (−V) will appear at $V_N$.

Figure 2.12 shows an analog multiplexer using MOSFETs. If the control input CON is HIGH (+$V_{CC}$), the MOSFET $Q_1$ is ON, $Q_2$ is OFF and (+V) will appear at $V_N$. If the control input CON is LOW (−$V_{CC}$), the MOSFET $Q_1$ is OFF, $Q_2$ is ON and (−V) will appear at $V_N$.

## 2.6 ASTABLE MULTIVIBRATOR

Fig. 2.13 shows an astable multivibrator using an op-amp. Let us assume that initially the op-amp output is LOW(i.e. negative saturation). The voltage at the non-inverting terminal will be

$$V_A = \beta(-V_{SAT})$$

$$\beta = \frac{R_1}{R_1 + R_2} \qquad (2.21)$$

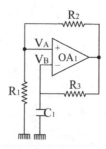

**FIGURE 2.13** Astable multivibrator

The voltage at the inverting terminal $V_B$ will be positive w.r.t. $V_A$ and its potential is decreasing, i.e. $C_1$ charges down through $R_3$. When the potential difference between the two input terminals approaches zero, the op-amp comes out of saturation. The positive feedback from the output to terminal $V_A$ causes regenerative switching, which drives the op-amp to positive saturation. The capacitor $C_1$ charges up through $R_3$ and $V_B$ potential rises exponentially; when it reaches $V_B = \beta(+Vcc)$ the circuit switches back to the state in which the op-amp is in negative saturation. The sequence therefore repeats to produce a square waveform of time period T at its output. The time period T is given as

$$T = 2R_3 C_1 \ln\left(1 + 2\frac{R_1}{R_2}\right) \tag{2.22}$$

Voltage to period converter: If in the astable multivibrator shown in Fig. 2.13, the $R_2$ terminal is removed from the output terminal and a controller is added between $R_2$ and the output as shown in Fig. 2.14, then the circuit will work as a voltage to time period converter. The time period T is given as

$$T = V_C K_1 \tag{2.23}$$

Where $K_1$ is constant depends on equation (2.22) and the op-amp saturation voltage or supply voltage Vcc.

Voltage to frequency converter: If in the astable multivibrator shown in Fig. 2.13, the $R_3$ terminal is removed from the output terminal and a controller is added between $R_3$ and the output as shown in Fig. 2.15, then the circuit will work as voltage to frequency converter. The frequency 'f' is given as

$$f = V_C K_1 \tag{2.24}$$

Where $K_1$ is constant depends on equation (2.22) and the op-amp saturation voltage or supply voltage Vcc.

Basic Components 17

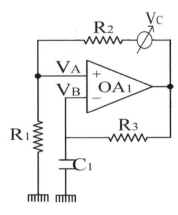

**FIGURE 2.14** Voltage to period converter

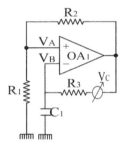

**FIGURE 2.15** Voltage to frequency converter

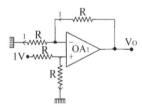

**FIGURE 2.16** Figure for question 2

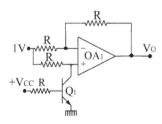

**FIGURE 2.17** Figure for question 3

## TUTORIAL EXERCISE

1. Design a suitable op-amp circuit to get (−5V) from the 1 V source.
2. Find the output voltage in the circuit shown in Fig. 2.16.
3. Find the output voltage in the circuit shown in Fig. 2.17.
4. For an integrator −1 V is applied at t = 0. Determine the time constant RC required such that the output reaches +10 V at t = 1 mS.

# 3 Non-linear Circuits

Linear circuits use (i) negative feedback to force the op-amp to operate within its linear region and (ii) linear elements with the feedback network. Non-linear circuits use a high-gain amplifier with positive feedback or sometimes no feedback at all, which causes op-amps to operate in saturation mode. Non-linear circuits use a feedback network with non-linear elements such as diodes and analog switches.

The non-linear circuits like comparator, Schmitt trigger, precision rectifiers, peak detectors, sample and hold circuits, log amplifiers and anti-log amplifiers are described in this chapter.

## 3.1 VOLTAGE COMPARATORS

Fig 3.1 shows the voltage comparator. It compares its positive input voltage $V_P$ and negative input voltage $V_N$ and produces output $V_O$.

The output voltage $V_O$ is

(i) HIGH ($+V_{SAT}$) when $V_P > V_N$
(ii) LOW ($-V_{SAT}$) when $V_P < V_N$

It is observed that $V_P$ and $V_N$ may be analog signals but output $V_O$ is a binary signal.

When both the inputs $V_P$ and $V_N$ are shorted together and a single voltage $V_D$ is applied, the output $V_0$ will be

(i) HIGH ($+V_{SAT}$) for $V_D > 0V$ or positive
(ii) LOW ($-V_{SAT}$) for $V_D < 0V$ or negative

The voltage transfer curve is shown in Fig 3.1(b). Fig 3.2(a) shows a comparator as a zero crossing detector.

In Fig 3.2(a) the comparator compares the input sine wave at its positive terminal and ground, which is at the inverting terminal. The output is (i) HIGH during the positive half-cycle of the input sine wave and (ii) LOW during the negative half-cycle of the input sine wave. A square waveform is generated at the output of the comparator and is shown in Fig. 3.2(b).

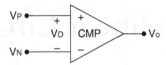

**FIGURE 3.1(A)**  Comparator

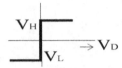

**FIGURE 3.1(B)**  Voltage transfer curve

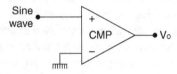

**FIGURE 3.2(A)**  Zero crossing detector (case I)

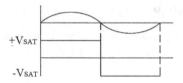

**FIGURE 3.2(B)**  Associated waveforms of 3.2(a)

In Fig 3.3(a) the comparator compares the input sine wave at its negative terminal and ground, which is at the non-inverting terminal. The output is (i) HIGH during the negative half-cycle of the input sine wave and (ii) LOW during the positive half-cycle of the input sine wave. A square waveform is generated at the output of the comparator and is shown in Fig. 3.3(b).

In Fig. 3.4(a) the comparator compares a sawtooth waveform $V_{SI}$ of peak value $V_R$ and time period T at its inverting terminal with an input voltage $V_1$ at its non-inverting terminal. A rectangular waveform is generated at the output of the comparator and is shown in Fig. 3.4(b). The HIGH or ON time of this rectangular waveform is given as

$$\delta_T = \frac{\text{Input voltage } V_1}{\text{Peak value of saw tooth}} \times \text{Time period} = \frac{V_1}{V_R} T \qquad (3.1)$$

Non-linear Circuits 21

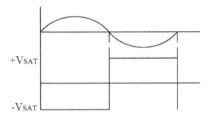

**FIGURE 3.3(A)** Zero crossing detector (case II)

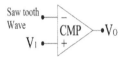

**FIGURE 3.3(B)** Associated waveforms of Fig. 3.3(a)

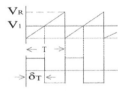

**FIGURE 3.4(A)** Comparator with sawtooth wave input (case I)

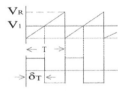

**FIGURE 3.4(B)** Associated waveforms of Fig. 3.4(a)

In Fig. 3.5(a) the comparator compares a sawtooth waveform of peak value $V_R$ and time period T at its non-inverting terminal with an input voltage $V_1$ at its inverting terminal. A rectangular waveform is generated at the output of the comparator and is shown in Fig. 3.5(b). The LOW or OFF time is given as

$$\delta_T = \frac{\text{Input voltage } V_1}{\text{Peak value of saw tooth}} \times \text{Time period} = \frac{V_1}{V_R} T \qquad (3.2)$$

In Fig. 3.6(a) the comparator compares a triangular waveform of $\pm V_T$ peak value and time period T at its inverting terminal with an input voltage $V_1$, which is at it non-inverting terminal. A rectangular waveform is generated at the output of the comparator and is shown in Fig. 3.6(b). The OFF time $T_1$ is given as

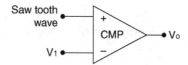

**FIGURE 3.5(A)** Comparator with sawtooth wave (case II)

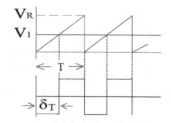

**FIGURE 3.5(B)** Associated waveform of Fig. 3.5(a)

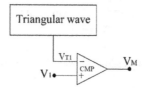

**FIGURE 3.6(A)** Comparison of triangular wave (case 1)

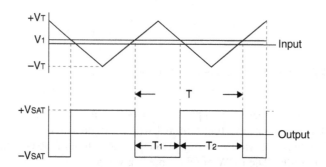

**FIGURE 3.6(B)** Associated waveforms of Fig. 3.6(a)

$$T_1 = \frac{\text{Peak value of triangular wave} - \text{Input voltage}}{\text{Twice Peak value of triangular wave}}$$

$$\times \text{Time period} = \frac{V_T - V_1}{2V_T} T \qquad (3.3)$$

# Non-linear Circuits

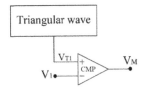

**FIGURE 3.7(A)** Comparison of triangular wave (case II)

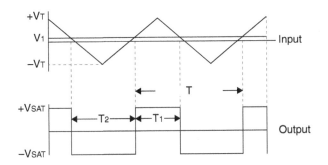

**FIGURE 3.7(B)** Associated waveforms of Fig. 3.7(a)

the ON time $T_2$ is given as

$$T_2 = \frac{\text{Peak value of triangular wave} + \text{Input voltage}}{\text{Twice Peak value of triangular wave}}$$
$$\times \text{Time period} = \frac{V_T + V_1}{2V_T} T \quad (3.4)$$

In Fig 3.7(a) the comparator compares a triangular waveform $V_{T1}$ of $\pm V_T$ peak value and time period T at its non-inverting terminal with an input voltage $V_1$, which is at its inverting terminal. A rectangular waveform is generated at the output of comparator and is shown in Fig. 3.7(b). The ON time $T_1$ is given as

$$T_1 = \frac{\text{Peak value of triangular wave} - \text{Input Voltage}}{\text{Twice Peak value of triangular wave}}$$
$$\times \text{Time period} = \frac{V_T - V_1}{2V_T} T \quad (3.5)$$

the OFF time $T_2$ is given as

$$T_2 = \frac{\text{Peak value of triangular wave} + \text{Input Voltage}}{\text{Twice Peak value of triangular wave}}$$
$$\times \text{Time period} = \frac{V_T + V_1}{2V_T} T \quad (3.6)$$

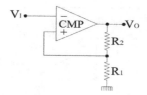

**FIGURE 3.8(A)**   Inverting Schmitt trigger

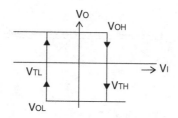

**FIGURE 3.8(B)**   Voltage transfer curve of Fig. 3.8(a)

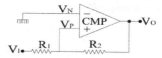

**FIGURE 3.9(A)**   Non-inverting Schmitt trigger

## 3.2   SCHMITT TRIGGERS

A Schmitt trigger is an amplifier with positive feedback. The negative feedback of an op-amp is used for an amplifier and the positive feedback of an op-amp is used for oscillation. The positive feedback makes the amplifier to go for saturation.

Fig. 3.8(a) shows circuit diagram of inverting Schmitt trigger.

The output has only two states, namely $V_{OH}$ and $V_{OL}$.

$$V_{TH} = \frac{R_1}{R_1 + R_2} V_{OH} \tag{3.7}$$

$$V_{TL} = \frac{R_1}{R_1 + R_2} V_{OL} \tag{3.8}$$

If the input voltage is less than zero (i.e. negative voltage) the op-amp goes to positive saturation (+ $V_{SAT}$). When the input voltage more than the $V_{TH}$ value, the op-amp output goes to negative saturation (−$V_{SAT}$).

Fig. 3.9 shows a non-inverting Schmitt trigger. For negative input voltage, the output will go to negative saturation

# Non-linear Circuits

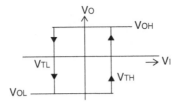

**FIGURE 3.9(B)**  Voltage transfer curve of Fig. 3.9(a)

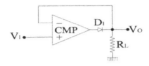

**FIGURE 3.10(A)**  Basic half-wave rectifier

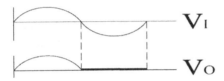

**FIGURE 3.10(B)**  Associated waveforms of Fig. 3.10(a)

$$V_{TH} = -\frac{R_1}{R_2} V_{OL} \qquad (3.9)$$

$$V_{TL} = -\frac{R_1}{R_2} V_{OH} \qquad (3.10)$$

## 3.3 HALF-WAVE RECTIFIERS

A half-wave rectifier is a circuit that passes only one portion of input wave (either positive or negative). The output of the positive half-wave rectifier is given as

$V_0 = V_I$ for positive input voltage
$V_0 = O$ for negative input voltage

Fig. 3.10(a) shows a basic half-wave rectifier. If the input voltage is positive, i.e. greater than zero volts, the op-amp output goes to positive saturation, the diode $D_1$ is forward biased and conducting, the op-amp will work as a buffer, $V_I$ appearing across the load resister $R_L$

$$V_o = +V_I \qquad (3.11)$$

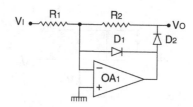

**FIGURE 3.11(A)** Half-wave rectifier

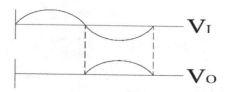

**FIGURE 3.11(B)** Associated waveforms of Fig. 3.11(a)

If the input voltage is less than zero volts or negative input voltage, the op-amp output goes to negative saturation, the diode $D_1$ is reverse biased and not conducting, and zero volts appears across the load resister $R_L$.

$$V_o = O \tag{3.12}$$

Fig. 3.11(a) shows an improved half-wave rectifier and its associated waveforms are shown in Fig. 3.11(b).

If the input voltage is greater than zero or for positive input voltage: Diode $D_1$ is ON, which creates a negative closed loop feedback, the inverting terminal voltage will be zero by virtual ground, op-amp output will be $-V_{D1}$, Diode $D_2$ is OFF, no current flows through $R_2$ and hence the output is zero voltage.

$$V_o = O \ [(\text{for } V_I > O) \tag{3.13}$$

If the input voltage is less than zero or for negative input voltage: The op-amp output goes to positive saturation, Diode $D_2$ is ON, diode $D_1$ is OFF, the circuit will work as inverting amplifier with a gain of $(R_2/R_1)$. The output voltage is given as

$$V_O = -V_I \left(\frac{R_2}{R_1}\right) [\text{for } V_I < O] \tag{3.14}$$

## 3.4 FULL-WAVE RECTIFIERS

A positive full-wave rectifier is a circuit that (i) passes the positive portion of the input voltage and (ii) inverts and then passes the negative portion of the input voltage. The output of a positive full-wave rectifier is given as

# Non-linear Circuits

$V_0 = +V_I$ for positive input voltage
$V_0 = -(-V_I)$ for negative input voltage

$$(\text{Or})\ V_0 = |V_I|$$

A full-wave rectifier is also known as an absolute-value circuit.

Fig. 3.12 shows the circuit diagram of a full-wave rectifier.

The op-amp $OA_1$ circuit constitutes a half-wave rectifier. During a positive half-cycle of input $V_I$, the diode $D_1$ is OFF and $D_2$ is ON, the op-amp $OA_1$ circuit will work as an inverting amplifier and $-V_I$ will appear at the junction $V_X$. During the negative half-cycle of input $V_I$, diode $D_1$ is ON and $D_2$ is OFF, and the op-amp $OA_1$ will be in a negative feedback loop configuration. Its non-inverting voltage will be equal to its inverting terminal voltage. Zero volts will appear at junction $V_X$. The voltage at $V_X$ is given as

$V_X = -V_I$ when $V_I > 0$
$V_X = 0$ when $V_I < 0$

The op-amp $OA_2$ circuit is an adder. It adds the voltages $V_X$ and $V_I$ and produces an output voltage $V_0$.

$$V_0 = -\left[\frac{R_5}{R_3}V_x + \frac{R_5}{R_4}V_I\right]$$

During positive half-cycle of the input $V_I$ i.e. when $V_I > 0$

$$V_0 = -\left[-V_I\left(\frac{R_5}{R_3}\right) + \frac{R_5}{R_4}V_I\right]$$

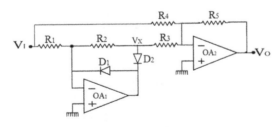

**FIGURE 3.12(A)**  Full wave rectifier

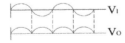

**FIGURE 3.12(B)**  Associated waveforms of Fig. 3.12(a)

Let $R_4 = R_5 = R$ and $R_3 = R/2$

$$V_0 = [2V_I - V_I] = V_I \qquad (3.15)$$

During a negative half-cycle of the input $V_I$, i.e. when $V_I < 0$

$$V_0 = -\left[\frac{R_5}{R_3}(0) + \frac{R_5}{R_4}V_I\right]$$

$$V_0 = -V_I \qquad (3.16)$$

The input and output waveforms are shown in Fig. 3.12(b)

Fig 3.13(a) shows another full-wave rectifier with two matched resistors.

During the positive half-cycle of the input $V_I$, i.e. $V_I > 0$, diode $D_1$ is ON, the op-amp $OA_1$ will be in negative closed loop configuration, its inverting terminal voltage will equal to its non-inverting terminal voltage, and zero volts will be the output of op-amp $OA_1$. Diode $D_2$ is OFF. The op-amp $OA_2$ will work as an non-inverting amplifier. Its output will be

$$V_0 = V_I\left(1 + \frac{R_3}{R_2}\right) \text{ when } V_I > 0 \qquad (3.17)$$

During the negative half-cycle of the input $V_I$, i.e. $V_I < 0$ diode $D_1$ is OFF, $D_2$ is on (forward biased by resistor $R_4$), op-amp $OA_1$, will still in negative closed loop via $D_2 - OA_2 - R_3 - R_2$. Hence its inverting terminal will be at zero volts by virtual ground.

By Kirchoss Law

$$I_1 = I_2$$

$$\frac{0 - V_I}{R_1} = \frac{V_0 - 0}{R_2 + R_3}$$

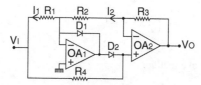

**FIGURE 3.13(A)** Full wave rectifier with two matched resistors

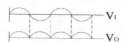

**FIGURE 3.13(B)** Associated waveforms of Fig. 3.13(a)

# Non-linear Circuits

$$-\frac{V_I}{R_1} = \frac{V_0}{R_2 + R_3}$$

$$\therefore V_0 = -V_I \frac{(R_2 + R_3)}{R_1}$$

Let us assume

$$1 + \frac{R_3}{R_2} = \frac{R_2 + R_3}{R_1}$$

$$\frac{R_2 + R_3}{R_2} = \frac{R_2 + R_3}{R_1}$$

Let $R_1 = R_2 = R$

$$V_0 = -V_I \left(1 + \frac{R_3}{R_2}\right) \text{ when } V_I < 0$$

The input and output waveforms are shown in Fig. 3.13(b).

## 3.5 PEAK DETECTORS

The peak detector detects the maximum value of a signal over a period of time. Fig. 3.14(a) shows a simple diode-capacitor peak detector.

The capacitor C is charged by the input signal through the diode. When the input signal falls, the diode is reverse biased and the capacitor voltage retains the peak

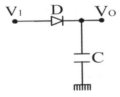

**FIGURE 3.14(A)** Simple diode-capacitor peak detector

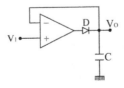

**FIGURE 3.14(B)** Op-amp peak detector

value of the input signal. This simple circuit has errors because of the diode forward voltage drop. This sort of forward voltage drop errors can be removed by replacing the diode with a precise diode as shown in Fig. 3.14(b). It operates in either a peak tracking mode or a peak storage mode. During peak tracking mode, the peak detector tracks the input towards a peak value. During peak storage mode, the peak detector holds this peak value as constant. Figs. 3.14(a),(b) are positive peak detectors and if we interchange the polarity of diode D, then it will become a negative peak detector.

If a sawtooth wave of peak value $V_p$ is given to a peak detector, the peak detector output is a DC voltage of $V_p$. Similarly, if a triangular wave of $\pm V_p$ value is given to a peak detector, if the peak detector is positive then its output will be a DC voltage of $+V_p$ and if the peak detector is negative then its output will be a DC voltage of $-V_p$.

## 3.6 SAMPLE AND HOLD CIRCUITS

The sample and hold circuit samples the input signal and holds it at a particular instant of time. This particular instant of time is determined by the sampling pulse. Fig. 3.15(a) shows a simple sample and hold circuit.

Let a sawtooth wave of peak value $V_p$ and time period T be given as the input to the sample and hold circuit. As shown in Fig. 3.15(a), a sampling pulse $V_s$ is given to the sample and hold circuit. The switch $S_1$ is closed during the HIGH time of sampling pulse $V_s$ and at that particular time of the input signal is given to the capacitor C and the capacitor C holds this signal even if the switch $S_1$ opens during the LOW time of the sampling pulse $V_s$. Hence the instant value of input signal is sampled and in hold with the sampling pulse $V_s$. This is illustrated in Fig 3.16. An op-amp buffer can be added at the output in order to avoid the loading effect, as shown in Fig. 3.15(b).

## 3.7 LOG AMPLIFIERS

The log amplifier using a diode is shown in Fig. 3.17.

$$I_R = \frac{V_I}{R_1} \qquad (3.18)$$

**FIGURE 3.15(A)**  Simple sample and hold circuit

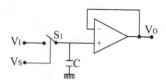

**FIGURE 3.15(B)**  Op-amp sample and hold circuit

# Non-linear Circuits

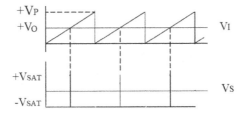

**FIGURE 3.16** Waveforms of Fig. 3.15(a), (b)

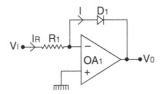

**FIGURE 3.17** Log amplifier with diode

The voltage across diode $V_D$ is given as

$$V_D = V_T \ln \frac{I}{I_S} \tag{3.19}$$

Where

$V_T = KT/q = 25$ mV at room temperature
$I$ = Current passing through diode
$I_s$ = Reverse saturation current

The inverting terminal of op-amp $OA_1$ is at the virtual ground and since it has high input impedance

$$I_R = I$$

The output voltage is given as

$$-V_D = -V_T \ln\left(\frac{I_R}{I_S}\right) = -V_T \ln\left(\frac{V_I}{R_1 I_S}\right) \tag{3.20}$$

The diode log amplifier is simple but has many drawbacks: (i) very poor log conformity and (ii) drifting of the output due to temperature variations. The drawbacks of diode log amplifiers are overcome by transistor log amplifiers.

Fig. 3.18 shows a transistor log amplifier.

**32**                                  Multiplier-Cum-Divider Circuits

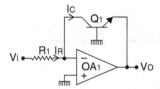

**FIGURE 3.18** Log amplifier with transistor

The logarithmic operation of bipolar transistor is given as

$$V_{BE} = V_T \ln \frac{I_C}{I_S} \tag{3.21}$$

Where $V_{BE}$ = Base emitter voltage

$V_T$ = KT/q = 25 mV at room temperature
$I_C$ = Collector current
$I_S$ = Emitter saturation current
$I_R = \dfrac{V_I}{R_1}$

The inverting terminal op-amp $OA_1$ is at virtual ground and since it has high input impedance

$$I_R = I_C$$

Output voltage $V_O$ is given as

$$V_0 = -V_{BE}$$
$$= -V_T \ln \left(\frac{I_C}{I_S}\right)$$

$$V_0 = -V_T \ln \left(\frac{V_I}{R_1 I_S}\right) \tag{3.22}$$

## 3.8 ANTI-LOG AMPLIFIERS

The anti-log amplifier using a diode is shown in Fig. 3.19. The inverting terminal of op-amp $OA_1$ will be at virtual ground and hence the voltage at its inverting terminal is zero volts. As the op-amp has high input impedance.

$$I_R = I_D$$

# Non-linear Circuits

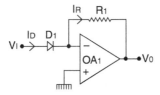

**FIGURE 3.19** Anti-log amplifier with diode

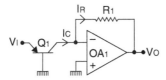

**FIGURE 3.20** Anti-log amplifier with transistor

The output voltage is given as

$$V_0 = -I_R R_1 = -I_D R_1$$

The diode current $I_D$ is given as

$$I_D = I_S e^{V_I/\eta V_T}$$

$$V_0 = -\left[I_S e^{V_I/\eta V_T}\right] R_1 = -(I_S R_1) e^{V_I/\eta V_T} \quad (3.23)$$

Let us assume $I_S R_1 = V_R$

$$V_O = -V_R e^{V_I/\eta V_T} \quad (3.24)$$

Thus, the output voltage is proportional to the exponential function of $V_I$. The exponential function is nothing but an anti-log function, and the circuit is working as an anti-log amplifier.

The anti-log amplifier using a single transistor is shown in Figure.3.20

The inverting terminal is op-amp $OA_1$ is at virtual ground, the inverting terminal voltage of op-amp $OA_1$ is zero volts. Since op-amp $OA_1$ has high input impedance,

$$I_C = I_R$$

$$I_C = I_S e^{V_{BE}/V_T} \quad (3.25)$$

The output voltage is given as

$$V_0 = -I_R R_1 = -I_C R_1$$

$$V_0 = I_S R_1 e^{V_{BE}/V_T}$$

Let $I_S R_1 = V_R$, $V_{BE} = V_I$ and hence

$$V_0 = V_R e^{V_I/V_T} \qquad (3.26)$$

Thus the output voltage is an exponential function of an input or an anti-log function of the input voltage.

## WORKED EXAMPLES

3.1 Fig. 3.21 shows a window detector using comparators. Explain its working operation.

The operation of the circuit is shown in Table 3.1. When $V_{IN}$ is less than $V_{TL}$, comparator $CMP_1$ output is HIGH, comparator $CMP_2$ output is LOW, diode $D_1$ is OFF, $D_2$ is ON and output $V_O$ will be LOW. When $V_{IN}$ is greater than $V_{TH}$, comparator $CMP_1$ output is LOW, comparator $CMP_2$ output is HIGH, diode $D_1$ is ON, diode $D_2$ is OFF, and output $V_O$ will be LOW. When $V_{TL} < V_{IN} < V_{TH}$, both comparators $CMP_1$ and

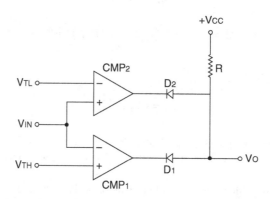

**FIGURE 3.21** Window detector

**TABLE 3.1**

| Input range | $CMP_1$ | $CMP_2$ | Output |
|---|---|---|---|
| $V_{IN} < V_{TL}$ | HIGH | LOW | LOW |
| $V_{TL} < V_{IN} < V_{TH}$ | HIGH | HIGH | HIGH |
| $V_{IN} > V_{TH}$ | LOW | HIGH | LOW |

# Non-linear Circuits

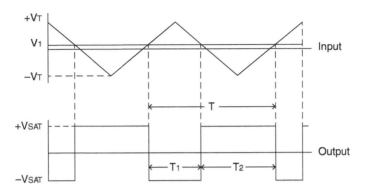

**FIGURE 3.22** Waveforms for problem 3.2

CMP$_2$ outputs are HIGH. Diodes D$_1$ and D$_2$ are OFF and the output voltage V$_o$ will be HIGH.

3.2 A triangular wave with ±10 Vpp and a frequency of 1 KHz is compared with a DC voltage of 1 V. The triangular wave is given to the inverted input of a comparator. Draw the output waveform of the comparator and find the ON and OFF times of this output waveform.

Given: V$_T$ = 5V, V$_1$ = 1V, T = 1/f = 1mS

$$T_1 = \frac{V_T - V_1}{2V_T} T = \frac{5-1}{10} 1mS = 0.4mS$$

$$T_2 = \frac{V_T + V_1}{2V_T} T = \frac{5+1}{10} 1mS = 0.6mS$$

The input and output waveforms are shown in Fig. 3.22.

# 4 Conventional MCDs

There are several conventional multiplier-cum-divider circuits (MCDs). MCDs using log–anti-log amplifiers, MCDs using FETs, and MCDs using MOSFETs are a few examples. These are discussed in this chapter.

## 4.1 LOG–ANTI-LOG MCDS – TYPE I

The log–anti-log multiplier type I is shown in Fig. 4.1. The logarithmic operation of a bipolar transistor is

$$V_{BE} = V_T \ln \frac{I_C}{I_S} \tag{4.1}$$

Where $V_{BE}$ = base emitter voltage, $V_T$ = KT/Q, $I_C$ = collector current and $I_S$ = emitter saturation current.

Considering log amp $OA_1$, the collector current of $Q_1$ will be the current through the resistor $R_1$.

$$I_{CQ1} = \frac{V_2}{R_1}$$

$$V_{BEQ1} = 0 - V_X = V_{TQ1} \ln \frac{I_{CQ1}}{I_{SQ1}} = V_{TQ1} \ln \frac{V_2}{R_1 I_{SQ1}} \tag{4.2}$$

Considering log amp $OA_3$, the collector current of $Q_3$ will be the current through the resistor $R_3$.

$$I_{CQ3} = \frac{V_1}{R_3}$$

37

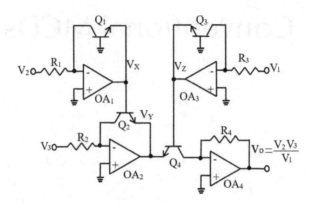

**FIGURE 4.1** Log–anti-log MCD – type I

$$V_{BEQ3} = 0 - V_Z = V_{TQ3} \ln \frac{I_{CQ3}}{I_{SQ3}} = V_{TQ3} \ln \frac{V_1}{R_3 I_{SQ3}} \quad (4.3)$$

Assume the transistors $Q_1$ and $Q_3$ are a matched pair such that $V_{TQ1} = V_{TQ3} = V_{T1}$ and $I_{SQ1} = I_{SQ3} = I_{S1}$, then the above equations (4.2) and (4.3) become

$$-V_X = V_{T1} \ln \frac{V_2}{R_1 I_{S1}} \quad (4.4)$$

$$-V_Z = V_{T1} \ln \frac{V_1}{R_3 I_{S1}} \quad (4.5)$$

Equations (4.4) and (4.5) give

$$-V_X + V_Z = V_{T1} \ln \frac{V_2}{R_1 I_{S1}} - V_{T1} \ln \frac{V_1}{R_3 I_{S1}}$$

$$V_Z - V_X = V_{T1} \ln \frac{V_2}{V_1} \frac{R_3}{R_1} \quad (4.6)$$

Consider the log amp $OA_2$; the collector current of $Q_2$ will be the current through the resistor $R_2$.

$$I_{CQ2} = \frac{V_3}{R_2}$$

$$V_{BEQ2} = V_X - V_Y = V_{TQ2} \ln \frac{I_{CQ2}}{I_{SQ2}} = V_{TQ2} \ln \frac{V_3}{R_2 I_{SQ2}} \quad (4.7)$$

# Conventional MCDs

Consider the anti-log amplifier $OA_4$; the collector current of $Q_4$ will be the current through the resistor $R_4$.

$$I_{CQ4} = \frac{V_O}{R_4}$$

$$V_{BEQ4} = V_Z - V_Y = V_{TQ4} \ln \frac{I_{CQ4}}{I_{SQ4}} = V_{TQ4} \ln \frac{V_O}{R_4 I_{SQ4}} \qquad (4.8)$$

Assume transistors $Q_2$ and $Q_4$ are matched such that $V_{TQ2} = V_{TQ4} = V_{T2}$ and $I_{SQ4} = I_{SQ2} = I_{S2}$.

$$V_X - V_Y = V_{T2} \ln \frac{V_3}{R_2 I_{S2}} \qquad (4.9)$$

$$V_Z - V_Y = V_{T2} \ln \frac{V_O}{R_4 I_{S2}} \qquad (4.10)$$

Equations (4.10) and (4.9) give

$$V_Z - V_Y - V_X + V_Y = V_{T2} \ln \frac{V_O}{R_4 I_{S2}} \frac{R_2 I_{S2}}{V_3}$$

$$V_Z - V_X = V_{T2} \ln \frac{V_O}{V_3} \frac{R_2}{R_4} \qquad (4.11)$$

Further assume that all transistors are matched and are kept adjacent to each other such that $V_{T2} = V_{T1} = V_T$, $I_{S2} = I_{S1} = I_S$ and assume $R_1 = R_2 = R_3 = R_4 = R$. Then equations (4.6) and (4.11)

$$V_{T1} \ln \frac{V_2}{V_1} \frac{R_3}{R_1} = V_{T2} \ln \frac{V_O}{V_2} \frac{R_2}{R_4} \qquad (4.12)$$

$$\ln \frac{V_2}{V_1} = \ln \frac{V_O}{V_3} \qquad (4.13)$$

$$V_O = \frac{V_2 V_3}{V_1} \qquad (4.14)$$

## 4.2 LOG–ANTI-LOG MCDS – TYPE II

The circuit diagram of log–anti-log MCD type II is shown in Fig. 4.2.
The logarithmic operation of bipolar transistor is

$$V_{BE} = V_T \ln \frac{I_C}{I_S} \qquad (4.15)$$

Where $V_{BE}$ = base emitter voltage, $V_T$ = KT/Q, $I_C$ = collector current and $I_S$ = emitter saturation current.

In Fig. 4.2 let us assume all transistors $Q_1$, $Q_2$, $Q_3$, and $Q_4$ are identical and packed in a single monolithic chip. Hence $V_{TQ1} = V_{TQ2} = V_{TQ3} = V_{TQ4} = V_T$ and $I_{SQ1} = I_{SQ2} = I_{SQ3} = I_{SQ4} = I_S$.

Logging transistor $Q_3$,

$$V_A - V_X = V_T \ln \frac{I_3}{I_S} \qquad (4.16)$$

Logging transistor $Q_1$

$$0 - V_X = V_T \ln \frac{I_1}{I_S} \qquad (4.17)$$

Equations (4.16) and (4.17) give

$$V_A = V_T \ln \frac{I_3}{I_1} \qquad (4.18)$$

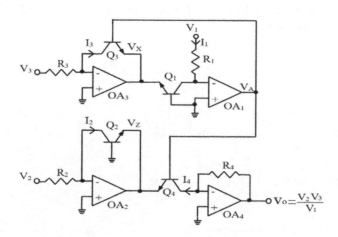

**FIGURE 4.2** Log–anti-log MCD – type II

Conventional MCDs

Logging transistor $Q_2$

$$0 - V_Z = V_T \ln \frac{I_2}{I_S} \tag{4.19}$$

Logging transistor $Q_4$

$$V_A - V_Z = V_T \ln \frac{I_4}{I_S} \tag{4.20}$$

Equations (4.20) and (4.19) give

$$V_A = V_T \ln \frac{I_4}{I_2} \tag{4.21}$$

From equations (4.18) and (4.21)

$$V_T \ln \frac{I_3}{I_1} = V_T \ln \frac{I_4}{I_2}$$

$$\frac{I_3}{I_1} = \frac{I_4}{I_2} \tag{4.22}$$

It is observed from the Fig. 4.2 that $I_1 = V_1/R_1$, $I_2 = V_2/R_2$, $I_3 = V_3/R_3$ and $I_4 = V_O/R_4$.

$$\frac{V_3 R_1}{R_3 V_1} = \frac{V_O R_2}{R_4 V_2} \tag{4.23}$$

Let us assume $R_1 = R_2 = R_3 = R_4 = R$, then

$$\frac{V_3}{V_1} = \frac{V_O}{V_2}$$

$$V_O = \frac{V_2 V_3}{V_1} \tag{4.24}$$

## 4.3  MCD USING FETS

The circuit diagram of MCD using FETs is shown in Fig. 4.3. FET is to be used in the triode region as VVR (Voltage Variable Resistor). The two FETs $Q_1$ and $Q_2$ are

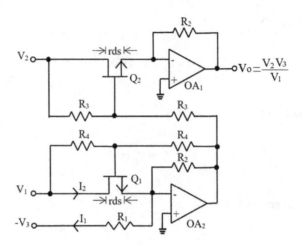

**FIGURE 4.3** Circuit diagram of FET-based multiplier

identical such that $V_{GS}$ and $r_{ds}$ of both FETs are the same. $R_3$ are linearizing resistors of $Q_2$ and $R_4$ are linearizing resistors of $Q_1$.

The output of inverting amplifier $OA_1$ will be

$$V_O = -\frac{R_2}{r_{ds}} V_2 \qquad (4.25)$$

The current through $R_1$ is given as

$$I_1 = -\frac{V_3}{R_1} \qquad (4.26)$$

The current through FET $Q_1$ is

$$I_2 = \frac{V_1}{r_{ds}}$$

$$I_1 = I_2$$

$$-\frac{V_3}{R_1} = \frac{V_1}{r_{ds}}$$

$$r_{ds} = -\frac{V_1 R_1}{V_3} \qquad (4.27)$$

# Conventional MCDs

Equation (4.27) in (4.25) gives

$$V_o = \frac{R_2 V_2 V_3}{V_1 R_1}$$

If $R_1 = R_2$, then

$$V_o = \frac{V_2 V_3}{V_1} \tag{4.28}$$

## 4.4 MCD USING MOSFETS

The MCD using MOSFETs is shown in Fig. 4.4. MOSFETs $Q_1$ and $Q_2$ are an identical matched pair and work as voltage controlled resistors. When the voltage between gate and drain exceeds the threshold voltage, i.e. $V_{GD} > V_{TH}$, the drain to source current is given as

$$I_{DS} = K[2(V_{GS} - V_T)V_{DS} - V_{DS}^2]$$

$$I_{DS} \simeq 2K(V_{GS} - V_T)V_{DS} \tag{4.29}$$

The input voltages $V_1$ and $V_2$ are positive and hence the source terminals of MOSFETs are at virtual ground by the op-amps $OA_1$ and $OA_2$. The current through $R_1$ will be

$$I_1 = -\frac{V_3}{R_1} \tag{4.30}$$

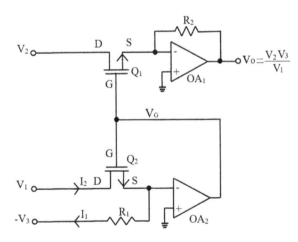

**FIGURE 4.4** MCD using MOSFETs

# Multiplier-Cum-Divider Circuits

Considering MOSFET $Q_2$

$$I_1 = I_2 = I_{DS2} = 2K(V_{GS2} - V_{T2})V_{DS2} \tag{4.31}$$

$$V_{GS2} = V_G \tag{4.32}$$

$$V_{DS2} = V_1 \tag{4.33}$$

Equations (4.32) and (4.33) in (4.31) give

$$I_1 = I_2 = 2K(V_G - V_{T2})V_1$$

$$(V_G - V_{T2}) = \frac{I_1}{2KV_1}$$

$$V_G - V_{T2} = -\frac{V_3}{2KR_1V_1} \tag{4.34}$$

Considering another MOSFET $Q_1$, we have

$$I_{DS1} = 2K(V_{GS1} - V_{T1})V_{DS1} \tag{4.35}$$

$$V_{GS1} = V_G \tag{4.36}$$

$$V_{DS1} = V_2 \tag{4.37}$$

$$V_{T1} = V_{T2}$$

Equations (4.36) and (4.37) in (4.35) give

$$I_{DS1} = 2K(V_G - V_{T1})V_2$$

$$I_{DS1} = -\frac{V_2 V_3}{R_1 V_1} \tag{4.38}$$

The output voltage $V_O$ will be

$$V_O = -I_{DS1} R_2 \tag{4.39}$$

Conventional MCDs

Let $R_1 = R_2 = R$. Equation (4.38) in (4.39) gives

$$V_O = \frac{V_2 V_3}{V_1} \quad (4.40)$$

## 4.5 MCD USING TWO ANALOG MULTIPLIERS

The block diagram of a MCD using two multipliers is shown in Fig. 4.5. The first multiplier $M_1$ output will be

$$V_A = \frac{V_2 V_3}{V_R} \quad (4.41)$$

The output of the second multiplier $M_2$ will be

$$V_B = \frac{V_1 V_O}{V_R} \quad (4.42)$$

In equations (4.41) and (4.42) $V_R$ are the multiplier constants. The op-amp OA is connected in a negative feedback loop and a positive DC voltage is ensured in the closed loop. Hence its inverting terminal voltage will be equal to its non-inverting terminal voltage, i.e. $V_A = V_B$.

$$\frac{V_2 V_3}{V_R} = \frac{V_1 V_O}{V_R} \quad (4.43)$$

$$V_O = \frac{V_2 V_3}{V_1} \quad (4.44)$$

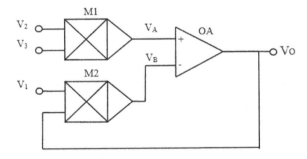

**FIGURE 4.5** Block diagram MCD using two multipliers and an op-amp

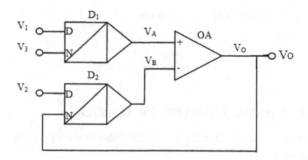

**FIGURE 4.6** MCD using two dividers

## 4.6 MCD USING TWO ANALOG DIVIDERS

The block diagram of the MCD using two dividers is shown in Fig. 4.6. The first divider $D_1$ output will be

$$V_A = \frac{N}{D} = \frac{V_3}{V_1} V_R \qquad (4.45)$$

The output of second divider $D_2$ will be

$$V_B = \frac{N}{D} = \frac{V_0}{V_2} V_R \qquad (4.46)$$

In equations (4.45) and (4.46) $V_R$ is the divider constant. The op-amp OA is configured in a negative closed loop configuration and a positive DC voltage is ensured in the feedback. Hence the voltage at its positive input voltage must be equal to the voltage at its negative input voltage

$$V_A = V_B \qquad (4.47)$$

$$\frac{V_3}{V_1} = \frac{V_0}{V_2}$$

$$V_O = \frac{V_2 V_3}{V_1} \qquad (4.48)$$

## 4.7 MCD USING TWO ANALOG DIVIDERS IN CASCADE

The block diagram of the MCD using two dividers is shown in Fig. 4.7. The first divider $D_1$ output will be

$$V_X = \frac{N}{D} = \frac{V_1}{V_2} V_R \qquad (4.49)$$

# Conventional MCDs

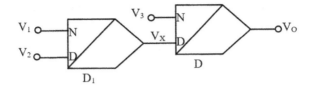

**FIGURE 4.7** Block diagram of an MCD using dividers

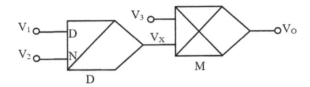

**FIGURE 4.8** Block diagram of MCD using divider and multiplier

The output of second divider $D_2$ will be

$$V_O = \frac{N}{D} = \frac{V_3}{V_X} V_R \qquad (4.50)$$

In equations (4.49) and (4.50) $V_R$ is the divider constant.

$$V_O = \frac{V_3 V_R}{V_1 V_R} V_2 \qquad (4.51)$$

$$V_O = \frac{V_2 V_3}{V_1} \qquad (4.52)$$

## 4.8 MCD USING A DIVIDER AND A MULTIPLIER IN SERIES

The block diagram of multiplier cum divider using a divider and a multiplier is shown in Fig. 4.8. The divider D output will be

$$V_X = \frac{N}{D} = \frac{V_2}{V_1} V_R \qquad (4.53)$$

The output of multiplier M will be

$$V_O = \frac{V_X V_3}{V_R} \qquad (4.54)$$

In equations (4.53) and (4.54) $V_R$ are the divider and multiplier constants.

$$V_O = \frac{V_2 V_R}{V_1 V_R} V_3 \qquad (4.55)$$

$$V_O = \frac{V_2 V_3}{V_1} \qquad (4.56)$$

# 5 Sawtooth Wave-Referenced Time-Division MCD Using Multiplexers

A pulse train whose maximum value is proportional to one voltage ($V_3$) is generated. If the width of this pulse train is made proportional to one voltage ($V_2$) and inversely proportional to another voltage ($V_1$), then the average value of pulse train is proportional to $\frac{V_2 V_3}{V_1}$. This is called a time-division multiplier-cum-divider (MCD). There are six types of time-division MCDs.

(i) Sawtooth wave-referenced MCD – multiplexing
(ii) Triangular wave-referenced MCD – multiplexing
(iii) MCD using no reference – multiplexing
(iv) Sawtooth wave-referenced MCD – switching
(v) Triangular wave-referenced MCD – switching
(vi) MCD using no reference – switching

The time-division MCD using sawtooth wave-referenced multiplexing is discussed in Section 5.2.

A pulse train whose OFF time is proportional to one voltage $V_2$ and inversely proportional to another voltage $V_1$ is generated. The third input voltage $V_3$ is integrated during this OFF time. The peak value of this integrated output is proportional to $\frac{V_2 V_3}{V_1}$. This is called a time-division peak-detecting MCD and is discussed in Section 5.3.

## 5.1 SAWTOOTH WAVE GENERATORS

Four circuits for the generation of a sawtooth wave are given in Figs. 5.1(a)–(d) and their associated waveforms in Fig. 5.2. A sawtooth wave $V_{S1}$ of peak value $V_R$ and time period T is generated by these circuits.

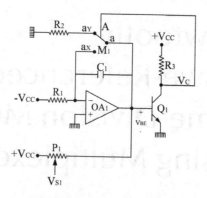

**FIGURE 5.1(A)**  Sawtooth wave generator – I

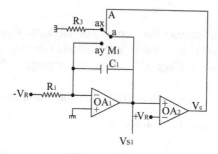

**FIGURE 5.1(B)**  Sawtooth wave generator – II

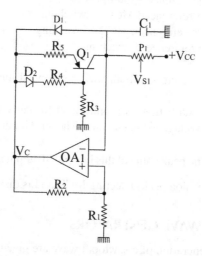

**FIGURE 5.1(C)**  Sawtooth wave generator – III

# Sawtooth Wave-Referenced Time-Division MCD

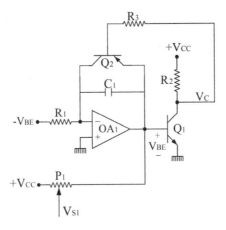

**FIGURE 5.1(D)** Sawtooth wave generator – IV

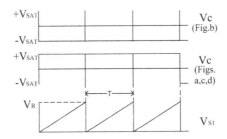

**FIGURE 5.2** Associated waveforms of Fig. 5.1

In Fig. 5.1(a),

$$V_R = 2V_{BE} \tag{5.1}$$

$$T = 1.4 R_1 C_1 \tag{5.2}$$

In Fig. 5.1(b), if initially op-amp $OA_2$ output is LOW, the multiplexer $M_1$ connects 'ax' to 'a' and the integrator formed by resister $R_1$, capacitor $C_1$ and op-amp $OA_1$ integrates $(-V_R)$ and its output is given as

$$V_{S1} = -\frac{1}{R_1 C_1} \int -V_R dt$$

$$V_{S1} = \frac{V_R}{R_1 C_1} t \tag{5.3}$$

A positive going ramp is generated at the output of op-amp $OA_1$ and when it reaches the value of reference voltage $+V_R$ the comparator $OA_2$ output becomes HIGH. The multiplexer $M_1$ now connects 'ay' to 'a' and shorts capacitor $C_1$ and hence integrator output becomes zero. Then comparator output is LOW and the sequence therefore repeats to give a perfect sawtooth wave $V_{S1}$ of peak value $V_R$ at the output of op-amp $OA_1$. From equation (5.3), Fig. 5.2 and fact that at t= T, $V_{S1} = V_R$ we can get

$$V_R = \frac{V_R}{R_1 C_1} T$$

$$T = R_1 C_1 \tag{5.4}$$

In Fig. 5.1(c), the time period T is given as

$$T = 2R_5 C_1 \ln\left(1 + 2\frac{R_1}{R_2}\right) \tag{5.5}$$

The peak value $V_R$ is given as

$$V_R = \beta(V_{SAT}) + \frac{\beta(V_{SAT})}{1.5} \tag{5.6}$$

Where $\beta$ is given as $\beta = \dfrac{R_1}{R_1 + R_2}$

In Fig. 5.1(d), the time period T is given as

$$T = 2R_1 C_1 \tag{5.7}$$

The peak value $V_R$ of this sawtooth $V_{S1}$ is given as

$$V_R = 2V_{BE} \tag{5.8}$$

Design exercise:

The sawtooth generator shown in Fig. 5.1(b) is modified as below: (i) The multiplexer ax, ay terminals are interchanged. (ii) comparator $OA_2$ +, − terminals are interchanged. (i) Draw circuit diagrams, (ii) draw waveforms at appropriate places, (iii) explain working operation, and (iv) deduce the expression for time period T.

## 5.2 DOUBLE MULTIPLEXING–AVERAGING TIME-DIVISION MCD

The circuit diagram of a double multiplexing–averaging time-division MCD is shown in Fig. 5.3 and its associated waveforms are shown in Fig. 5.4. As discussed in the

# Sawtooth Wave-Referenced Time-Division MCD

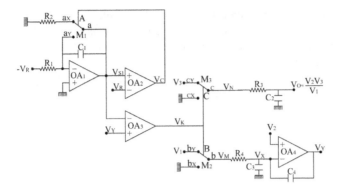

**FIGURE 5.3** Double multiplexing–averaging MCD

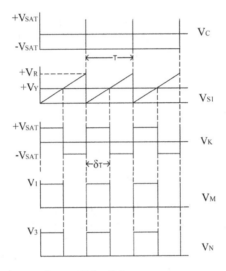

**FIGURE 5.4** Associated waveforms of Fig. 5.3

previous section, a sawtooth wave $V_{S1}$ of peak value $V_R$ and time period T is generated by the op-amps $OA_1$, $OA_2$, and multiplexer $M_1$. The time period T is given as

$$T = R_1 C_1 \tag{5.9}$$

The comparator $OA_3$ compares the sawtooth wave $V_{S1}$ with the voltage $V_Y$ and produces a rectangular waveform $V_K$. The ON time $\delta_T$ of $V_K$ is given as

$$\delta_T = \frac{V_Y}{V_R} T \tag{5.10}$$

The rectangular pulse $V_K$ controls the second multiplexer $M_2$. When $V_K$ is HIGH, the first input voltage $V_1$ is connected to the $R_4C_3$ low pass filter ('by' is connected to 'b'). When $V_K$ is LOW, zero volts are connected to the $R_4C_3$ low pass filter ('bx' is connected to 'b'). Another rectangular pulse $V_M$ with a maximum value of $V_1$ is generated at the multiplexer $M_2$ output. The $R_4C_3$ low pass filter gives an average value of this pulse train $V_M$ and is given as

$$V_X = \frac{1}{T}\int_0^{\delta_T} V_1 dt = \frac{V_1}{T}\delta_T \tag{5.11}$$

$$V_X = \frac{V_1 V_Y}{V_R} \tag{5.12}$$

The op-amp $OA_4$ is configured in a negative closed loop feedback and a positive DC voltage is ensured in the feedback loop. Hence its inverting terminal voltage must be equal to its non-inverting terminal voltage.

$$V_2 = V_X \tag{5.13}$$

From equations (5.12) and (5.13)

$$V_Y = \frac{V_2 V_R}{V_1} \tag{5.14}$$

The rectangular pulse $V_K$ also controls the third multiplexer $M_3$. When $V_K$ is HIGH, the third input voltage $V_3$ is connected to the $R_3C_2$ low pass filter ('cy' is connected to 'c'). When $V_K$ is LOW, zero volts are connected to the $R_3C_2$ low pass filter ('cx' is connected to 'c'). Another rectangular pulse $V_N$ with maximum value of $V_3$ is generated at the multiplexer $M_3$ output. The $R_3C_2$ low pass filter gives an average value of this pulse train $V_N$ and is given as

$$V_O = \frac{1}{T}\int_0^{\delta_T} V_3 dt = \frac{V_3}{T}\delta_T \tag{5.15}$$

Equations (5.10) and (5.14) in (5.15) give

$$V_O = \frac{V_2 V_3}{V_1} \tag{5.16}$$

Design exercise:

1. In the MCD circuit shown in Fig. 5.3, the sawtooth wave generator shown in Fig. 5.1(b) is used. Replace this sawtooth wave generator with other sawtooth

generators shown in Fig. 5.1(a), 5.1(c) and 5.1(d). In each (i) draw circuit diagrams, (ii) draw waveforms at appropriate places, (iii) explain working operation, and (iv) deduce the expression for their output voltages.
2. Replace multiplexer $M_2$ and $M_3$ in Fig 5.3 with a transistor multiplexer, FET multiplexer, and MOSFET multiplexer. In each, (i) draw circuit diagrams, (ii) draw waveforms at appropriate places, (iii) explain the working operation, and (iv) deduce an expression for their output voltages.

## 5.3 TIME-DIVISION SINGLE-SLOPE PEAK-DETECTING MCD

The circuit diagram of time-division single-slope peak-detecting MCD is shown in Fig. 5.5 and its associated waveforms in Fig. 5.6. As discussed in Section 5.1, a sawtooth wave $V_{S1}$ with peak value $V_R$ and time period T is generated by op-amp $OA_1$, transistor $Q_1$, and multiplexer $M_1$.

$$T = 1.4 R_1 C_1$$

$$V_R = 2V_{BE} \quad (5.17)$$

The comparator $OA_2$ compares the sawtooth wave $V_{S1}$ with an input voltage $V_Y$ and produces a rectangular waveform $V_M$. The OFF time $\delta_T$ of $V_M$ is given as

$$\delta_T = \frac{V_Y}{V_R} T \quad (5.18)$$

The rectangular pulse $V_M$ controls the second multiplexer $M_2$. When $V_M$ is HIGH, the input voltage $V_1$ is connected to the $R_6 C_3$ low pass filter ('by' is connected to 'b'). When $V_M$ is LOW, zero volts is connected to the $R_6 C_3$ low pass filter ('bx' is connected to 'b'). Another rectangular pulse $V_N$ with a maximum value of $V_1$ is generated at the

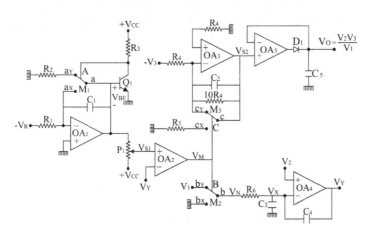

**FIGURE 5.5** Sawtooth wave-based time-division multiplier type – I

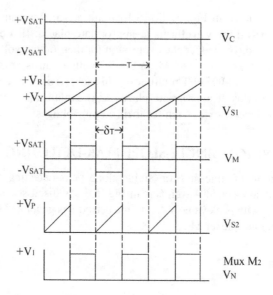

**FIGURE 5.6** Associated waveforms of Fig. 5.5

multiplexer $M_2$ output. The $R_6C_3$ low pass filter gives an average value of this pulse train $V_N$ and is given as

$$V_X = \frac{1}{T}\int_0^{\delta_T} V_1 dt = \frac{V_1}{T}\delta_T \qquad (5.19)$$

$$V_X = \frac{V_1 V_Y}{V_R} \qquad (5.20)$$

The op-amp $OA_4$ is configured in a negative closed loop feedback and a positive DC voltage is ensured in the feedback loop. Hence its inverting terminal voltage must be equal to its non-inverting terminal voltage.

$$V_2 = V_X \qquad (5.21)$$

From equations (5.20) and (5.21)

$$V_Y = \frac{V_2}{V_1}V_R \qquad (5.22)$$

The rectangular pulse $V_M$ also controls the multiplexer $M_3$. During HIGH of $V_M$, multiplexer $M_3$ selects 'cy' to 'c,' the capacitor $C_2$ is short circuited so that the op-amp $OA_3$ output is zero volts. During LOW of $V_M$, the multiplexer $M_3$ selects 'cx' to

# Sawtooth Wave-Referenced Time-Division MCD

'c,' the integrator formed by resistor $R_4$, capacitor $C_2$, op-amp $OA_3$ integrates its input voltage $(-V_3)$, and its output is given as

$$V_{S2} = -\frac{1}{R_4 C_2}\int -V_3 dt = \frac{V_3}{R_4 C_2} t \qquad (5.23)$$

A semi sawtooth waveform $V_{S2}$ with peak value $V_P$ is generated at the output of op-amp $OA_3$. From the waveforms shown in Fig. 5.6 and from equation (5.23), the fact that at $t = \delta_T$, $V_{S2} = V_P$.

$$V_P = \frac{V_3}{R_4 C_2} \delta_T \qquad (5.24)$$

Equations (5.18) and (5.22) in (5.24)

$$V_P = \frac{V_2 V_3}{V_1} \frac{T}{R_4 C_2}$$

Let us assume $T = R_4 C_2$

$$V_P = \frac{V_2 V_3}{V_1} \qquad (5.25)$$

The peak detector realized by op-amp $OA_5$, diode $D_1$ and capacitor $C_5$ gives this peak value $V_P$ at its output $V_O$. $V_O = V_P$. Hence the output will be

$$V_O = \frac{V_2 V_3}{V_1} \qquad (5.26)$$

Design exercise:

1. In the MCD circuit shown in Fig. 5.5, a sawtooth wave generator of Fig. 5.1(a) is used. Replace this sawtooth wave generator with other sawtooth wave generators shown in Figs. 5.1(b),(c),(d). In each (i) draw circuit diagrams, (ii) draw waveforms at appropriate places, (iii) explain working operation, and (iv) deduce the expression for their output voltages
2. Replace multiplexer $M_2$ in Fig 5.5 with a transistor multiplexer, FET multiplexer and MOSFET multiplexer. In each, (i) draw circuit diagrams, (ii) draw waveforms at appropriate places, (iii) explain working operation, and (iv) deduce expression for their output voltages.
3. In the MCD circuit shown in Fig. 5.5, (i) the op-amp $OA_2$ terminals (ii) 'by,' 'bx' terminals of multiplexer $M_2$, are interchanged. (i) Draw circuit diagrams, (ii) draw waveforms at appropriate places, (iii) explain working operation, and (iv) deduce the expression for their output voltages.

4. For the MCD circuit designed from question (3), a sawtooth wave generator of Fig. 5.1(a) is used. Replace this sawtooth wave generator with other sawtooth wave generators shown in Figs. 5.1(b), (c), and (d). In each (i) draw circuit diagrams, (ii) draw waveforms at appropriate places, (iii) explain working operation, and (iv) deduce the expression for their output voltages.
5. Replace multiplexer $M_2$ in the MCD circuit designed from question (3) with a transistor multiplexer, FET multiplexer and MOSFET multiplexer. In each, (i) draw circuit diagrams, (ii) draw waveforms at appropriate places, (iii) explain working operation, and (iv) deduce expression for their output voltages.

## 5.4 TIME-DIVISION MULTIPLY–DIVIDE MCD

The circuit diagram of sawtooth wave-based time-division multiply–divide MCD is shown in Fig. 5.7 and its associated waveforms are shown in Fig. 5.8. As discussed in Section 5.1, a sawtooth wave $V_{S1}$ with peak value $V_R$ and time period T is generated by op-amp $OA_3$, transistor $Q_1$, and multiplexer $M_1$.

$$T = 1.4 R_1 C_1 \tag{5.27}$$

$$V_R = 2 V_{BE}$$

The comparator $OA_2$ compares the sawtooth wave $V_{S1}$ with an input voltage $V_1$ and produces a rectangular waveform $V_M$. The ON time $\delta_{T1}$ of $V_M$, is given as

$$\delta_{T1} = \frac{V_3}{V_R} T \tag{5.28}$$

The rectangular pulse $V_M$ controls the second multiplexer $M_2$. When $V_M$ is HIGH, the second input voltage $V_2$ is connected to the $R_4 C_2$ low pass filter ('by' is connected

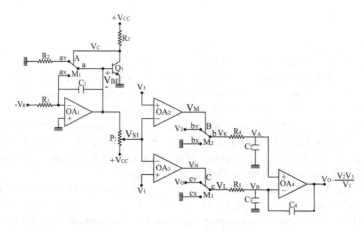

**FIGURE 5.7** Sawtooth wave-based time-division multiplier type – I

# Sawtooth Wave-Referenced Time-Division MCD

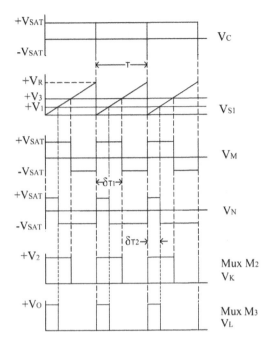

**FIGURE 5.8** Associated waveforms of Fig.5.7

to 'b'). When $V_M$ is LOW, zero volts are connected to the $R_4C_2$ low pass filter ('bx' is connected to 'b'). Another rectangular pulse $V_K$ with maximum value of $V_2$ is generated at the multiplexer $M_2$ output. The $R_4C_2$ low pass filter gives an average value of this pulse train $V_K$ and is given as

$$V_A = \frac{1}{T} \int_0^{\delta_{T1}} V_2 \, dt = \frac{V_2}{T} \delta_{T1} \qquad (5.29)$$

$$V_A = \frac{V_2 V_3}{V_R} \qquad (5.30)$$

The comparator $OA_3$ compares the sawtooth wave $V_{S1}$ with the first input voltage $V_1$ and produces a rectangular waveform $V_N$. The ON time $\delta_{T2}$ of $V_N$ is given as

$$\delta_{T2} = \frac{V_1}{V_R} T \qquad (5.31)$$

The rectangular pulse $V_N$ controls the third multiplexer $M_3$. When $V_N$ is HIGH, the output voltage $V_O$ is connected to $R_5C_3$ low pass filter ('cy' is connected to 'c'). When $V_N$ is LOW, zero volts is connected to $R_5C_3$ low pass filter ('cx' is connected to 'c'). Another rectangular pulse $V_L$ with a maximum value of $V_O$ is generated at the multiplexer $M_3$ output. The $R_5C_3$ low pass filter gives an average value of this pulse train $V_L$ and is given as

$$V_B = \frac{1}{T}\int_0^{\delta_{T2}} V_O dt = \frac{V_O}{T}\delta_{T2}$$

$$V_B = \frac{V_1 V_O}{V_R} \tag{5.32}$$

The op-amp $OA_4$ is configured in a negative closed loop feedback and a positive DC voltage is ensured in the feedback loop. Hence its inverting terminal voltage must equal to its non-inverting terminal voltage, i.e.

$$V_A = V_B \tag{5.33}$$

Equations (5.30) and (5.32) in (5.33) give

$$V_O = \frac{V_2 V_3}{V_1} \tag{5.34}$$

Design exercise:

1. In the MCD circuit shown in Fig. 5.7, a sawtooth wave generator shown in Fig. 5.1(a) is used. Replace this sawtooth wave generator with other sawtooth generators of Figs. 5.1(b), (c), and (d). In each (i) draw circuit diagrams, (ii) draw waveforms at appropriate places, (iii) explain its working operation, and (iv) deduce the expression for their output voltages.
2. Replace multiplexer $M_2$ and $M_3$ in Fig 5.7 with a transistor multiplexer, FET multiplexer and MOSFET multiplexer. In each, (i) draw circuit diagrams, (ii) draw waveforms at appropriate places, (iii) explain its working operation, and (iv) deduce the expression for their output voltages.
3. In the MCD circuit shown in Fig. 5.7, (i) the op-amp $OA_2$ and op-amp $OA_3$, and (ii) multiplexer $M_2$ and multiplexer $M_3$ are interchanged, (i) Draw circuit diagrams, (ii) draw waveforms at appropriate places, (iii) explain its working operation, and (iv) deduce the expression for their output voltages.
4. In the MCD circuit designed from above question (3), a sawtooth wave generator shown in Fig. 5.1(a) is used. Replace this sawtooth wave generator with other sawtooth generators of Figs. 5.1(b), (c), and (d). In each (i) draw circuit diagrams, (ii) draw waveforms at appropriate places, (iii) explain its working operation, and (iv) deduce the expression for their output voltages.
5. Replace the multiplexers $M_2$ and $M_3$ in the MCD circuit designed from question (3) with a transistor multiplexer of FET multiplexer and MOSFET multiplexer. In each, (i) draw circuit diagrams, (ii) draw waveforms at

appropriate places, (iii) explain its working operation, and (iv) deduce the expression for their output voltages.

## 5.5 TIME-DIVISION DIVIDE–MULTIPLY MCD

The circuit diagram of sawtooth wave-based time-division divide–multiply MCD is shown in Fig. 5.9 and its associated waveforms are shown in Fig. 5.10. As discussed in Section 5.1, a sawtooth wave $V_{S1}$ with peak value $V_{BE}$ and time period T is generated by op-amp $OA_3$, transistor $Q_1$, and multiplexer $M_1$.

$$T = 1.4 R_1 C_1 \tag{5.35}$$

$$V_R = 2V_{BE} \tag{5.36}$$

The comparator $OA_2$ compares the sawtooth wave with an input voltage $V_1$ and produces a rectangular waveform $V_M$. The ON time $\delta_{T1}$ of $V_M$ is given as

$$\delta_{T1} = \frac{V_1}{V_R} T \tag{5.37}$$

The rectangular pulse $V_M$ controls the second multiplexer $M_2$. When $V_M$ is HIGH, the voltage $V_Y$ is connected to the $R_4 C_2$ low pass filter ('by' is connected to 'b'). When $V_M$ is LOW zero volts is connected to the $R_4 C_2$ low pass filter ('bx' is connected to 'b'). Another rectangular pulse $V_K$ with a maximum value of $V_Y$ is generated at the multiplexer $M_2$ output. The $R_4 C_2$ low pass filter gives an average value of this pulse train $V_K$ and is given as

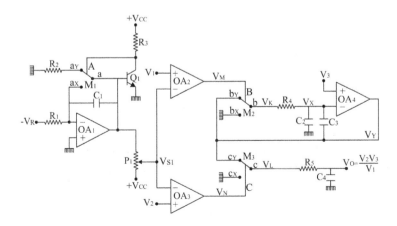

**FIGURE 5.9** Sawtooth wave-based time-division multiplier type – I

## Multiplier-Cum-Divider Circuits

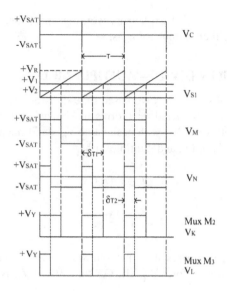

**FIGURE 5.10** Associated waveforms of Fig. 5.9

$$V_X = \frac{1}{T}\int_0^{\delta_{T1}} V_Y dt = \frac{V_Y}{T}\delta_{T1}$$

$$V_X = \frac{V_1 V_Y}{V_R} \quad (5.38)$$

The op-amp $OA_4$ is configured in a negative closed loop feedback and a positive DC voltage is ensured in the feedback loop. Hence its inverting terminal voltage must equal to its non-inverting terminal voltage, i.e.

$$V_X = V_3 \quad (5.39)$$

From equations (5.38) and (5.39)

$$V_Y = \frac{V_3 V_R}{V_1} \quad (5.40)$$

The comparator $OA_3$ compares the sawtooth wave with the second input voltage $V_2$ and produces a rectangular waveform $V_N$. The ON time $\delta_{T2}$ of $V_N$ is given as

# Sawtooth Wave-Referenced Time-Division MCD

$$\delta_{T2} = \frac{V_2}{V_R} T \tag{5.41}$$

The rectangular pulse $V_N$ controls the third multiplexer $M_3$. When $V_N$ is HIGH, the voltage $V_Y$ is connected to $R_5C_4$ low pass filter ('cy' is connected to 'c'). When $V_N$ is LOW, zero volts are connected to the $R_5C_4$ low pass filter ('cx' is connected to 'c'). Another rectangular pulse $V_L$ with a maximum value of $V_Y$ is generated at the multiplexer $M_3$ output. The $R_5C_4$ low pass filter gives an average value of this pulse train $V_L$ and is given as

$$V_O = \frac{1}{T} \int_0^{\delta_{T2}} V_Y dt = \frac{V_Y}{T} \delta_{T2} \tag{5.42}$$

$$V_O = \frac{V_2 V_Y}{V_R} \tag{5.43}$$

Equation (5.40) in (5.43) gives

$$V_O = \frac{V_2 V_3}{V_1} \tag{5.44}$$

Design exercise:

1. In the MCD circuit shown in Fig. 5.9, a sawtooth wave generator of Fig. 5.1(a) is used. Replace this sawtooth wave generator with other sawtooth generators of Fig. 5.1(b), (c), and (d). In each, (i) draw circuit diagrams, (ii) draw waveforms at appropriate places, (iii) explain its working operation, and (iv) deduce the expression for their output voltages.
2. Replace multiplexers $M_2$ and $M_3$ in Fig 5.9 with a transistor multiplexer FET multiplexer and MOSFET multiplexer. In each, (i) draw circuit diagrams, (ii) draw waveforms at appropriate places, (iii) explain its working operation, and (iv) deduce the expression for their output voltages.
3. In the MCD circuit shown in Fig. 5.9, (i) the op-amp $OA_2$ and op-amp $OA_3$, and (ii) multiplexer $M_2$ and multiplexer $M_3$ are interchanged. (i) Draw circuit diagrams, (ii) draw waveforms at appropriate places, (iii) explain its working operation, and (iv) deduce the expression for their output voltages.
4. In the MCD circuit designed from above question (3), a sawtooth wave generator shown in Fig. 5.1(a) is used. Replace this sawtooth wave generator with other sawtooth generators of Figs. 5.1(b), (c), and (d). In each (i) draw circuit diagrams, (ii) draw waveforms at appropriate places, (iii) explain its working operation, and (iv) deduce the expression for their output voltages.

5. Replace multiplexers $M_2$ and $M_3$ in the MCD circuit designed from question (3) with a transistor multiplexer of FET multiplexer and MOSFET multiplexer. In each, (i) draw circuit diagrams, (ii) draw waveforms at appropriate places, (iii) explain its working operation, and (iv) deduce the expression for their output voltages.

# 6 Triangular Wave-Referenced MCD with Multiplexers

A pulse train whose maximum value is proportional to one voltage ($V_3$) is generated. If the width of this pulse train is made proportional to one voltage ($V_2$) and inversely proportional to another voltage ($V_1$), then the average value of pulse train is proportional to $\dfrac{V_2 V_3}{V_1}$. This is called a time-division multiplier-cum-divider (MCD). There are six types of time-division MCDs.

(i) Sawtooth wave-referenced MCD – multiplexing
(ii) Triangular wave-referenced MCD – multiplexing
(iii) MCD using no reference – multiplexing
(iv) Sawtooth wave-referenced MCD – switching
(v) Triangular wave-referenced MCD – switching
(vi) MCD using no reference – switching

The time-division MCD using triangular wave-referenced multiplexing is discussed in Section 6.2.

A pulse train whose OFF time is proportional to one voltage $V_2$ and inversely proportional to another voltage $V_1$ is generated. The third input voltage $V_3$ is integrated during this OFF time. The peak value of this integrated output is proportional to $\dfrac{V_2 V_3}{V_1}$. This is called a time-division peak-detecting MCD and is discussed in Section 6.3.

## 6.1 TRIANGULAR WAVE GENERATORS

A triangular wave $V_{T1}$ with a $\pm V_T$ peak-to-peak value and time period T is generated by the triangular wave generators shown in Figs. 6.1(a),(b) and their associated waveforms are shown in Fig. 6.2.

# Multiplier-Cum-Divider Circuits

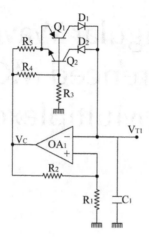

**FIGURE 6.1(A)** Triangular wave generator – I

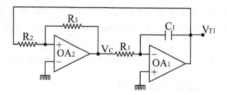

**FIGURE 6.1(B)** Triangular wave generator – II

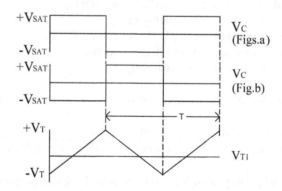

**FIGURE 6.2** Associated waveforms of Fig. 6.1

In Fig. 6.1(a)

$$V_T = \beta(V_{SAT}) \tag{6.1}$$

$$T = 4R_5 C_1 \frac{R_1}{R_2} \tag{6.2}$$

# Triangular Wave-Referenced MCD

Where $\beta$ is given as $\beta = \dfrac{R_1}{R_1 + R_2}$

In Fig. 6.1(b), op-amps $OA_1$ and $OA_2$ constitute a triangular/square wave generator. The output of op-amp $OA_1$ is a triangular wave $V_{T1}$ with $\pm V_T$ peak values and a time period of T. Initially the comparator $OA_2$ output is LOW($-V_{SAT}$); the output of the integrator composed by op-amp $OA_1$, resistor $R_1$ and capacitor $C_1$, is given as

$$V_{T1} = -\frac{1}{R_1 C_1} \int -V_{SAT} dt = \frac{V_{SAT}}{R_1 C_1} t \tag{6.3}$$

The integrator output is rising towards positive saturation and when it reaches a value $+V_T$, the comparator output becomes HIGH($+V_{SAT}$). The output of the integrator composed by op-amp $OA_1$, resistor $R_1$ and capacitor $C_1$, is given as

$$V_{T1} = -\frac{1}{R_1 C_1} \int +V_{SAT} dt = -\frac{V_{SAT}}{R_1 C_1} t$$

Now the output of the integrator changes its slope from $+V_T$ towards $(-V_T)$ and when it reaches a value '$-V_T$,' the comparator output becomes LOW($-V_{SAT}$) and the sequence therefore repeats to give (i) a triangular waveform $V_{T1}$ with $\pm V_T$ peak-to-peak values at the output of op-amp $OA_1$ and (ii) a square waveform $V_C$ with $\pm V_{SAT}$ peak-to-peak values at the output of comparator $OA_2$.

From the waveforms shown in Fig. 6.2, from equation (6.3) and the fact that at $t = T/2$, $V_{T1} = 2V_T$

$$2V_T = \frac{V_{SAT}}{R_1 C_1} \frac{T}{2}$$

$$T = \frac{4 V_T R_1 C_1}{V_{SAT}} \tag{6.4}$$

When the comparator $OA_2$ output is LOW $(-V_{SAT})$, the effective voltage at non-inverting terminal of comparator $OA_2$ will be by the superposition principle

$$\frac{(-V_{SAT})}{(R_2 + R_3)} R_2 + \frac{(+V_T)}{(R_2 + R_3)} R_3$$

When this effective voltage at the non-inverting terminal of comparator $OA_2$ becomes zero

$$\frac{(-V_{SAT}) R_2 + (+V_T) R_3}{(R_2 + R_3)} = 0$$

$$(+V_T) = (+V_{SAT})\frac{R_2}{R_3}$$

When the comparator $OA_2$ output is HIGH $(+V_{SAT})$, the effective voltage at the non-inverting terminal of comparator $OA_2$ will be by the superposition principle

$$\frac{(+V_{SAT})}{(R_2+R_3)}R_2 + \frac{(-V_T)}{(R_2+R_3)}R_3$$

When this effective voltage at non-inverting terminal of comparator $OA_2$ becomes zero

$$\frac{(+V_{SAT})R_2 + (-V_T)R_3}{(R_2+R_3)} = 0$$

$$(-V_T) = (-V_{SAT})\frac{R_2}{R_3}$$

$$\pm V_T = \pm V_{SAT}\frac{R_2}{R_3} \approx 0.76(\pm V_{CC})\frac{R_2}{R_3} \qquad (6.5)$$

From equation (6.4) and (6.5), the time period T of the generated triangular/square waveforms is given by

$$T = 4R_1C_1\frac{R_2}{R_3} \qquad (6.6)$$

## 6.2 TIME-DIVISION MCD

The circuit diagram of the triangular wave-referenced time-division MCD is shown in Fig. 6.3 and its associated waveforms are shown in Fig. 6.4. A triangular wave $V_{T1}$

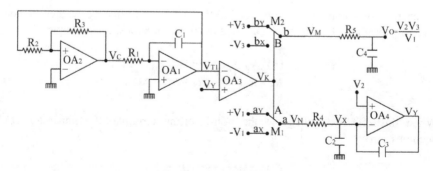

**FIGURE 6.3** Triangular wave-based time-division MCD

# Triangular Wave-Referenced MCD

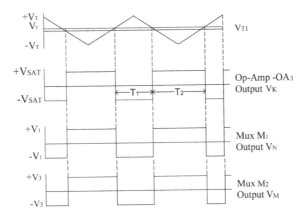

**FIGURE 6.4** Associated waveforms of Fig. 6.3

of $\pm V_T$ peak-to-peak values and the time period T is generated by the op-amps $OA_1$ and $OA_2$. The comparator $OA_3$ compares the triangular wave $V_{T1}$ with the voltage $V_Y$ and produce the asymmetrical rectangular wave $V_K$. From Fig 6.4, it is observed that

$$T_1 = \frac{V_T - V_Y}{2V_T} T$$

$$T_2 = \frac{V_T + V_Y}{2V_T} T$$

$$T = T_1 + T_2 \tag{6.7}$$

The rectangular wave $V_K$ controls multiplexer $M_1$, which connects $(+V_1)$ to its output during $T_2$ ('ay' is connected to 'a') and $(-V_1)$ to its output during $T_1$ ('ax' is connected to 'a'). Another asymmetrical rectangular waveform $V_N$ is generated at the multiplexer $M_1$ output with $\pm V_1$ peak-to-peak values. The $R_4C_2$ low pass filter gives the average value of $V_N$ and is given as

$$V_X = \frac{1}{T}\left[\int_0^{T_2} V_1\,dt + \int_{T_2}^{T_1+T_2}(-V_1)\,dt\right] = \frac{V_1}{T}[T_2 - T_1]$$

$$V_X = \frac{V_1 V_Y}{V_T} \tag{6.8}$$

The op-amp $OA_4$ is configured in a negative closed loop feedback and a positive DC voltage is ensured in the feedback loop. Hence its inverting terminal voltage must equal to its non-inverting terminal voltage, i.e.

$$V_X = V_2 \qquad (6.9)$$

From equations (6.8) and (6.9)

$$V_Y = \frac{V_2 V_T}{V_1} \qquad (6.10)$$

The rectangular wave $V_K$ also controls multiplexer $M_2$, which connects $(+V_3)$ to its output during $T_2$ ('by' is connected to 'b') $(-V_3)$ to the output during $T_1$ ('bx' is connected to 'b'). Another asymmetrical rectangular wave $V_M$ is generated at the multiplexer $M_2$ output with $\pm V_3$ peak-to-peak values. The $R_5 C_4$ low pass filter gives the average value $V_O$ and is given as

$$V_O = \frac{1}{T}\left[\int_0^{T_2} V_3\, dt + \int_{T_2}^{T_1+T_2}(-V_3)\,dt\right] = \frac{V_3}{T}[T_2 - T_1]$$

$$V_O = \frac{V_3 V_Y}{V_T} \qquad (6.11)$$

Equation (6.10) in (6.11) gives

$$V_O = \frac{V_2 V_3}{V_1} \qquad (6.12)$$

Design exercise:

1. In the MCD circuit shown in Fig. 6.3, the triangular wave generator shown in Fig. 6.1(b) is used. Replace this triangular wave generator with the other triangular wave generator shown in Fig. 6.1(a). (i) Draw circuit diagrams, (ii) draw waveforms at appropriate places, (iii) explain the working operation, and (iv) deduce the expression for their output voltages
2. Replace multiplexers $M_1$ and $M_2$ in Fig 6.3 with a transistor multiplexer, FET multiplexer and MOSFET multiplexer. In each, (i) draw circuit diagrams, (ii) draw waveforms at appropriate places, (iii) explain the working operation, and (iv) deduce the expression for their output voltages

## 6.3 TIME-DIVISION DIVIDE–MULTIPLY MCD

The circuit diagram of a triangular wave-based divide–multiply MCD is shown in Fig. 6.5 and its associated waveforms are shown in Fig. 6.6. A triangular wave $V_{T1}$ of $\pm V_T$ peak-to-peak values and the time period T is generated by the op-amp $OA_1$.

# Triangular Wave-Referenced MCD

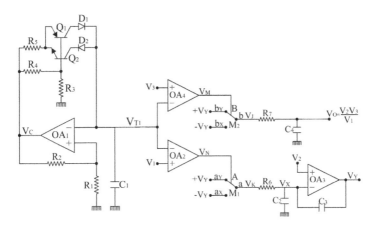

**FIGURE 6.5** Triangular wave-based divide–multiply MCD

The first input voltage $V_1$ is compared with the triangular wave $V_{T1}$ by the comparator on OA$_2$. An asymmetrical rectangular waveform $V_N$ is generated at the comparator OA$_2$ output. From the waveforms shown in Fig. 6.6, it is observed that

$$T_1 = \frac{V_T - V_1}{2V_T}T \quad T_2 = \frac{V_T + V_1}{2V_T}T \quad T = T_1 + T_2 \qquad (6.13)$$

This rectangular wave $V_N$ is given as the control input to the multiplexer M$_1$. The multiplexer M$_1$ connects the voltage (+$V_Y$) during $T_2$ ('ay' is connected to 'a') and (–$V_Y$) during $T_1$ ('ax' is connected to 'a'). Another rectangular asymmetrical wave $V_K$ with peak-to-peak values of ±$V_Y$ is generated at the multiplexer M$_1$ output. The $R_6C_2$ low pass filter gives the average value of the pulse train $V_N$, which is given as

$$V_X = \frac{1}{T}\left[\int_0^{T_2} V_Y\, dt + \int_{T_2}^{T_1+T_2}(-V_Y)\,dt\right] = \frac{V_Y}{T}(T_2 - T_1)$$

$$V_X = \frac{V_1 V_Y}{V_T} \qquad (6.14)$$

The op-amp OA$_3$ is configured in a negative closed loop feedback and a positive DC voltage is ensured in the feedback loop. Hence its inverting terminal voltage must be equal to its non-inverting terminal voltage, i.e.

$$V_2 = V_X \qquad (6.15)$$

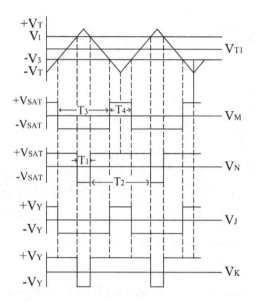

**FIGURE 6.6** Associated waveforms of Fig. 6.5

From equations (6.14) and (6.15)

$$V_Y = \frac{V_2 V_T}{V_1} \tag{6.16}$$

The third input voltage $V_3$ is compared with the generated triangular wave $V_{T1}$ by the comparator on $OA_4$. An asymmetrical rectangular waveform $V_M$ is generated at the comparator $OA_4$ output. From the waveforms shown in Fig. 6.6, it is observed that

$$T_3 = \frac{V_T - V_3}{2V_T} T \quad T_4 = \frac{V_T + V_3}{2V_T} T \quad T = T_3 + T_4 \tag{6.17}$$

This rectangular wave $V_M$ is given as the control input to the multiplexer $M_2$. The multiplexer $M_2$ connects the voltage $(+V_Y)$ during $T_4$ ('by' is connected to 'b') and $(-V_Y)$ during $T_3$ ('bx' is connected to 'b'). Another rectangular asymmetrical wave $V_J$ with peak-to-peak values of $\pm V_Y$ is generated at the multiplexer $M_2$ output. The $R_7 C_4$ low pass filter gives the average value of the pulse train $V_J$ and is given as

$$V_O = \frac{1}{T}\left[\int_0^{T_4} V_Y \, dt + \int_{T_4}^{T_3+T_4} (-V_Y) \, dt \right] = \frac{V_Y}{T}(T_4 - T_3) \tag{6.18}$$

# Triangular Wave-Referenced MCD

$$V_O = \frac{V_Y V_3}{V_T} \tag{6.19}$$

Equation (6.16) in (6.19) gives

$$V_O = \frac{V_2 V_3}{V_1} \tag{6.20}$$

Design exercise:

1. In the MCD circuit shown in Fig. 6.5, the triangular wave generator shown in Fig. 6.1(a) is used. Replace this triangular wave generator with other triangular wave generators shown in Fig. 6.1(b). (i) Draw circuit diagrams, (ii) draw waveforms at appropriate places, (iii) explain working operation, and (iv) deduce the expression for their output voltages.
2. Replace multiplexers $M_1$ and $M_2$ in Fig 6.5 with a transistor multiplexer, FET multiplexer and MOSFET multiplexer. In each, (i) draw circuit diagrams, (ii) draw waveforms at appropriate places, (iii) explain working operation, and (iv) deduce the expression for their output voltages.

## 6.4 TIME-DIVISION MULTIPLY–DIVIDE MCD

The circuit diagrams of triangular wave-based multiply–divide time-division MCDs is shown in Fig. 6.7 and their associated waveforms are shown in Fig. 6.8. A triangular wave of $\pm V_T$ peak-to-peak value and the time period T is generated around op-amp $OA_1$.

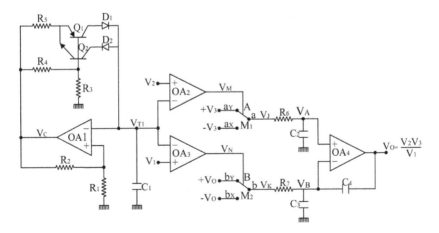

**FIGURE 6.7** Triangular wave-based time-division multiply–divide MCD

The second input voltage $V_2$ is compared with the triangular wave $V_{T1}$ by the comparator on $OA_2$. An asymmetrical rectangular waveform $V_M$ is generated at the comparator $OA_2$ output. From the waveforms shown in Fig. 6.8, it is observed that

$$T_1 = \frac{V_T - V_2}{2V_T} T, \quad T_2 = \frac{V_T + V_2}{2V_T} T, \quad T = T_1 + T_2 \qquad (6.21)$$

This rectangular wave $V_M$ is given as the control input to the multiplexer $M_1$. The multiplexer $M_1$ connects the third input voltage $(+V_3)$ during $T_2$ ('ay' is connected to 'a') and $(-V_3)$ during $T_1$ ('ax' is connected to 'a'). Another rectangular asymmetrical square wave $V_J$ with peak-to-peak values of $\pm V_3$ is generated at the multiplexer $M_1$ output. The $R_6 C_2$ low pass filter gives the average value of the pulse train $V_J$ and is given as

$$V_A = \frac{1}{T} \left[ \int_0^{T_2} V_3 \, dt + \int_{T_2}^{T_1 + T_2} (-V_3) \, dt \right] = \frac{V_3}{T}(T_2 - T_1)$$

$$V_A = \frac{V_2 V_3}{V_T} \qquad (6.22)$$

The first input voltage $V_1$ is compared with the generated triangular wave $V_{T1}$ by the comparator on $OA_3$. An asymmetrical rectangular waveform $V_N$ is generated at the comparator $OA_3$ output. From the waveforms shown in Fig. 6.8, it is observed that

$$T_3 = \frac{V_T - V_1}{2V_T} T, \quad T_4 = \frac{V_T + V_1}{2V_T} T, \quad T = T_3 + T_4 \qquad (6.23)$$

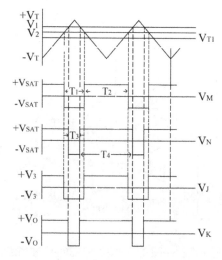

**FIGURE 6.8** Associated waveforms of Fig. 6.7

# Triangular Wave-Referenced MCD

This rectangular wave $V_N$ is given as the control input to the multiplexer $M_2$. The multiplexer $M_2$ connects the output voltage $(+V_O)$ during $T_4$ ('by' is connected to 'b') and $(-V_O)$ during $T_3$ ('bx' is connected to 'b'). Another rectangular asymmetrical wave $V_K$ with peak-to-peak values of $\pm V_O$ is generated at the multiplexer $M_2$ output. The $R_7C_3$ low pass filter gives the average value of the pulse train $V_N$ and is given as

$$V_B = \frac{1}{T}\left[\int_0^{T_4} V_O\, dt + \int_{T_4}^{T_3+T_4} (-V_O)\, dt\right] = \frac{V_O}{T}(T_4 - T_3)$$

$$V_B = \frac{V_1 V_O}{V_T} \qquad (6.24)$$

The op-amp $OA_4$ is configured in a negative closed loop feedback and a positive DC voltage is ensured in the feedback loop. Hence its inverting terminal voltage must equal to its non-inverting terminal voltage, i.e.

$$V_A = V_B \qquad (6.25)$$

From equations (6.22) and (6.24)

$$V_O = \frac{V_2 V_3}{V_1} \qquad (6.26)$$

Design exercise:

1. In the MCD circuit shown in Fig. 6.7, the triangular wave generator shown in Fig. 6.1(a) is used. Replace this triangular wave generator with the other triangular wave generator shown in Fig. 6.1(b). (i) Draw circuit diagrams, (ii) draw waveform at appropriate places, (iii) explain working operation, and (iv) deduce the expression for their output voltages.
2. Replace multiplexers $M_1$ and $M_2$ in Fig. 6.7 with a transistor multiplexer, FET multiplexer and MOSFET multiplexer. In each, (i) draw circuit diagrams, (ii) draw waveforms at appropriate places, (iii) explain working operation, and (iv) deduce the expression for their output voltages.

# 7 Peak-Responding MCDs with Multiplexers

Peak-responding MCDs are classified as (i) peak-detecting MCDs and (ii) peak-sampling MCDs. A short pulse / sawtooth waveform whose time period T is (i) proportional to one voltage ($V_2$) and (ii) inversely proportional to another voltage ($V_1$) is generated. The third input voltage $V_3$ is integrated during this time period T. The peak value of the integrated output is proportional to $\frac{V_2 V_3}{V_1}$. This is called a double single-slope peak-responding MCD. A square wave / triangular waveform whose time period T is (i) proportional to one voltage ($V_2$) and (ii) inversely proportional to one voltage ($V_1$) is generated. The third input voltage $V_3$ is integrated during this time period T. The peak value of the integrated output is proportional to $\frac{V_2 V_3}{V_1}$. This is called a double dual-slope peak-responding MCD. A rectangular waveform whose (i) time period T is inversely proportional to the first input voltage $V_1$ and (ii) OFF time is proportional to the second input voltage ($V_2$), is generated. The third input voltage $V_3$ is integrated during this OFF time. The peak value of the integrated output is proportional to $\frac{V_2 V_3}{V_1}$. This is called a pulse-width integrated peak-responding MCD.

At the output of a peak-responding MCD, if a peak detector is used, it is called a peak-detecting MCD and if sample and hold is used, it is called a peak-sampling MCD. The peak-responding MCDs of both peak-detecting MCDs and peak-sampling MCDs using analog two-to-one multiplexers are described in this chapter.

## 7.1 DOUBLE SINGLE-SLOPE PEAK-RESPONDING MCD

The circuit diagrams of double single-slope peak-responding MCDs are shown in Fig. 7.1 and their associated waveforms are shown in Fig. 7.2. Fig. 7.1(a) shows a double single-slope peak-detecting MCD and Fig. 7.1(b) shows a double single-slope peak-sampling MCD. Let's initially assume the comparator $OA_2$ output is LOW, the

# Multiplier-Cum-Divider Circuits

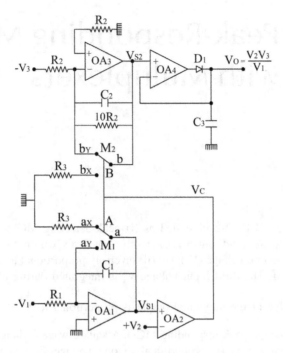

**FIGURE 7.1(A)** Double single-slope peak-detecting MCD

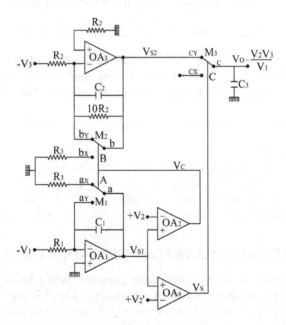

**FIGURE 7.1(B)** Double single-slope peak-sampling MCD

# Peak-Responding MCDs with Multiplexers

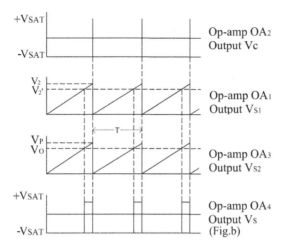

**FIGURE 7.2** Associated waveforms of Fig. 7.1

multiplexer $M_1$ connects 'ax' to 'a,' an integrator formed by resistor $R_1$, capacitor $C_1$ and op-amp $OA_1$ integrates the first input voltage $(-V_1)$. The integrated output will be

$$V_{S1} = -\frac{1}{R_1 C_1}\int -V_1 dt = \frac{V_1}{R_1 C_1}t \tag{7.1}$$

A positive going ramp $Vs_1$ is generated at the output of op-amp $OA_1$. When the output of $OA_1$ reaches the voltage level of $V_2$, the comparator $OA_2$ output becomes HIGH. The multiplexer $M_1$ connects 'ay' to 'a' and hence the capacitor $C_1$ is shorted so that the op-amp $OA_1$ output becomes ZERO. Then the comparator $OA_2$ output goes to LOW, the multiplexer $M_1$ connects 'ax' to 'a' and the integrator composed by $R_1$, $C_1$ and op-amp $OA_1$ integrates the input voltage $(-V_1)$ and the cycle therefore repeats to provide (i) a sawtooth wave of peak value $V_2$ at the output of op-amp $OA_1$ and (ii) a short pulse waveform $V_C$ at the output of comparator $OA_2$. The short pulse $V_C$ also controls the multiplexer $M_2$. During the short HIGH time of $V_C$, multiplexer $M_2$ selects 'by' to 'b,' the capacitor $C_2$ is short circuited so that the op-amp $OA_3$ output is zero volts. During the LOW time of $V_C$, the multiplexer $M_2$ selects 'bx' to 'b,' the integrator formed by resistor $R_2$, capacitor $C_2$, op-amp $OA_3$ integrates its input voltage $(-V_3)$ and its output is given as

$$V_{S2} = -\frac{1}{R_2 C_2}\int -V_3 dt = \frac{V_3}{R_2 C_2}t \tag{7.2}$$

Another sawtooth waveform $V_{S2}$ with peak value $V_P$ is generated at the output of op-amp $OA_3$. From the waveforms shown in Fig. 7.2 and from equations (7.1), (7.2) the fact that at $t = T$, $V_{S1} = V_2$, $V_{S2} = V_P$ we get

$$V_2 = \frac{V_1}{R_1 C_1} T \tag{7.3}$$

$$V_P = \frac{V_3}{R_2 C_2} T \tag{7.4}$$

From equations (7.3) and (7.4)

$$V_P = \frac{V_3}{R_2 C_2} \frac{V_2}{V_1} R_1 C_1$$

Let us assume $R_1 = R_2$, $C_1 = C_2$, then

$$V_P = \frac{V_2 V_3}{V_1} \tag{7.5}$$

(i) In Fig. 7.1(a), the peak detector realized by op-amp $OA_4$, diode $D_1$ and capacitor $C_3$ gives this peak value $V_P$ at its output $V_O$. $V_O = V_P$.

(ii) In Fig. 7.1(b), the peak value $V_P$ of the sawtooth waveform $V_{S2}$ is obtained by the sample and hold circuit realized by multiplexer $M_3$ and capacitor $C_3$. The sampling pulse is generated by the op-amp $OA_4$ by comparing a voltage slightly smaller than that of $V_2$, called $V_2'$ with the sawtooth wave $V_{S1}$. The sample and hold operation is illustrated graphically in Fig. 7.2. The sample and hold output $V_O$ is equal to $V_P$.

Hence the output will be $V_O = V_P$

$$V_O = \frac{V_2 V_3}{V_1} \tag{7.6}$$

Design exercise:

1. In Fig. 7.1(a) and (b), if polarity of $-V_3$ and direction of diode $D_1$ are reversed, draw the waveforms at the appropriate places and deduce the expression for output voltage.
2. In Figs. 7.1(a) and (b), if the input terminals of op-amp $OA_2$ and multiplexers $M_1$ and $M_2$ are interchanged, (i) draw a circuit diagram, (ii) explain the working operation, (iii) draw waveforms at appropriate places, and (iv) deduce the expression for the output voltage.

## 7.2 DOUBLE DUAL-SLOPE PEAK-RESPONDING MCDs USING FEEDBACK COMPARATOR

The circuit diagrams of double dual-slope peak-responding MCDs using a feedback comparator are shown in Figs. 7.3 and their associated waveforms are shown in Fig 7.4. Fig. 7.3(a) shows a peak-detecting MCD and Fig. 7.3(b) shows a peak-sampling MCD. Let's initially assume the comparator $OA_2$ output is LOW. The multiplexer $M_1$

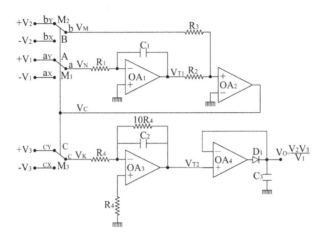

**FIGURE 7.3(A)** Double dual-slope peak-detecting MCD using a feedback comparator

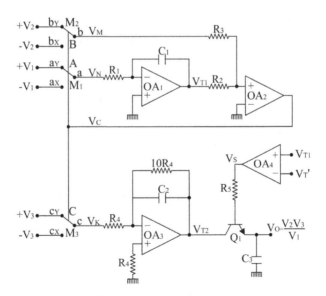

**FIGURE 7.3(B)** Double dual-slope peak- sampling MCD using a feedback comparator

selects $(-V_1)$ to the integrator I composed by resistor $R_1$, capacitor $C_1$ and op-amp $OA_1$ ('ax' is connected to 'a'). The integrator I output is given as

$$V_{T1} = -\frac{1}{R_1C_1}\int(-V_1)dt = \frac{V_1}{R_1C_1}t \qquad (7.7)$$

The output of integrator I is going towards positive saturation and when it reaches a value '$+V_T$,' the comparator $OA_2$ output becomes HIGH. The multiplexer $M_1$ selects $(+V_1)$ to the integrator I composed by resistor $R_1$, capacitor $C_1$ and op-amp $OA_1$ ('ay' is connected to 'a'). The integrator I output is given as

$$V_{T1} = -\frac{1}{R_1C_1}\int(+V_1)dt = -\frac{V_1}{R_1C_1}t \qquad (7.8)$$

The output of integrator I is reversing towards negative saturation and when it reaches a value '$(-V_T)$,' the comparator $OA_2$ output becomes LOW. The multiplexer $M_1$ selects $(-V_1)$ and the sequence repeats to give (i) a triangular waveform $V_{T1}$ of $\pm V_T$ peak-to-peak values with a time period of 'T' and (ii) square waveform $V_C$ at the output of comparator $OA_2$. From the waveforms shown in Fig.7.4, equation (7.7) and the fact that at $t = T/2$, $V_{T1} = 2V_T$

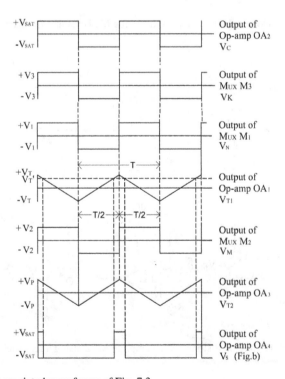

**FIGURE 7.4** Associated waveforms of Fig. 7.3

# Peak-Responding MCDs with Multiplexers

$$2V_T = \frac{V_1}{R_1 C_1} \frac{T}{2}$$

$$T = \frac{4R_1 C_1 V_T}{V_1} \qquad (7.9)$$

When the comparator $OA_2$ output is LOW $(-V_{SAT})$, $(-V_2)$ will be at the output of multiplexer $M_2$ ('bx' is connected to 'b'), and the effective voltage at non-inverting terminal of comparator $OA_2$ will be by superposition principle

$$\frac{(-V_2)}{(R_2 + R_3)} R_2 + \frac{(+V_T)}{(R_2 + R_3)} R_3$$

When this effective voltage at non-inverting terminal of comparator $OA_2$ becomes zero

$$\frac{(-V_2)R_2 + (+V_T)R_3}{(R_2 + R_3)} = 0$$

$$(+V_T) = (+V_2)\frac{R_2}{R_3}$$

When the comparator $OA_2$ output is HIGH $(+V_{SAT})$, $(+V_2)$ will be at the output of multiplexer $M_2$ ('by' is connected to 'b'), and the effective voltage at non-inverting terminal of comparator $OA_2$ will be by superposition principle

$$\frac{(+V_2)}{(R_2 + R_3)} R_2 + \frac{(-V_T)}{(R_2 + R_3)} R_3$$

When this effective voltage at non-inverting terminal of comparator $OA_2$ becomes zero

$$\frac{(+V_2)R_2 + (-V_T)R_3}{(R_2 + R_3)} = 0$$

$$(-V_T) = (-V_2)\frac{R_2}{R_3}$$

$$\pm V_T = \pm V_2 \frac{R_2}{R_3} \qquad (7.10)$$

The multiplexer $M_3$ connects $(+V_3)$ during the HIGH of the square waveform $V_C$ ('cy' is connected to 'c') and $-V_3$ during the LOW of $V_C$ ('cx' is connected to 'c'). Another square waveform $V_K$ with a $\pm V_3$ peak-to-peak value is generated at the output of multiplexer $M_3$. This square wave $V_K$ is converted into a triangular wave $V_{T2}$ by the integrator II composed by resistor $R_4$, capacitor $C_2$ and op-amp $OA_3$ with $\pm V_p$ as peak-to-peak values of the same time period T. For one transition the integrator II output is given as

$$V_{T2} = -\frac{1}{R_4 C_2} \int (-V_3) dt = \frac{V_3}{R_4 C_2} t \qquad (7.11)$$

From the waveforms shown in Fig. 7.4, the equation (7.11) and the fact that at $t = T/2$, $V_{T2} = 2V_p$ we get

$$2V_p = \frac{V_3}{R_4 C_2} \frac{T}{2} \qquad (7.12)$$

Equations (7.9) and (7.10) in (7.12) give

$$V_p = \frac{V_2 V_3}{V_1} \frac{R_1 C_1}{R_4 C_2} \frac{R_2}{R_3}$$

Let $R_1 = R_4$ and $C_1 = C_2$

$$V_p = \frac{V_2 V_3}{V_1} \frac{R_2}{R_3}$$

Let $R_2/R_3 = 1$ (but in practical $R_2/R_3 < 1$).

$$V_p = \frac{V_2 V_3}{V_1} \qquad (7.13)$$

(i) In Fig. 7.3(a), the peak detector at the output stage gives peak value '$V_p$' of triangular wave $V_{T2}$ and hence $V_0 = V_p$.
(ii) In Fig. 7.3(b), the peak value $V_p$ of the triangular waveform $V_{T2}$ is obtained by the sample and hold circuit realized by transistor $Q_1$ and capacitor $C_3$. The sampling pulse VS is generated by op-amp OA4 by comparing a slightly smaller voltage that that of $V_T$ called $V_T'$ with the triangular wave of $V_{T1}$. The sample and hold operation is illustrated graphically in Fig. 7.4. The sampled output is given as $V_0 = V_p$.

Peak-Responding MCDs with Multiplexers

From equation (7.13), $V_O = V_P$

$$V_O = \frac{V_2 V_3}{V_1} \quad (7.14)$$

Design exercise:

The multiplexers $M_1$, $M_2$, and $M_3$ in Figs. 7.3(a) and (b) are to be replaced with transistor multiplexers, FET multiplexers and MOSFET multiplexers. In each, (i) draw the circuit diagrams, (ii) explain their working operation, (iii) draw waveforms at appropriate places, and (iv) deduce the expression for the output voltage.

## 7.3 DOUBLE DUAL-SLOPE PEAK-RESPONDING MCD WITH FLIP FLOP

The circuit diagram of double dual-slope peak-responding MCDs is shown in Fig 7.5 and their associated waveforms are shown in Fig 7.6. Fig. 7.5(a) shows a peak-detecting MCD and 7.5(b) shows a peak-sampling MCD. Let initially the flip flop output be LOW. The multiplexer $M_1$ selects $(-V_1)$ to the integrator I composed by resistor $R_1$, capacitor $C_1$, and op-amp $OA_1$ ('ax' is connected to 'a'). The integrator I output is given as

$$V_{T1} = -\frac{1}{R_1 C_1} \int (-V_1) dt = \frac{V_1}{R_1 C_1} t \quad (7.15)$$

The output of integrator I is going towards positive saturation and when it reaches the value '$+V_2$,' the comparator $OA_2$ output becomes HIGH and it sets the flip flop output

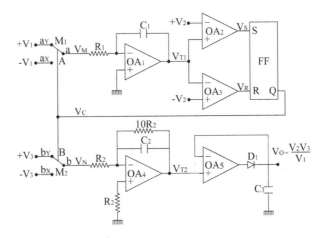

**FIGURE 7.5(A)** Double dual-slope peak-detecting MCD with flip flop

# Multiplier-Cum-Divider Circuits

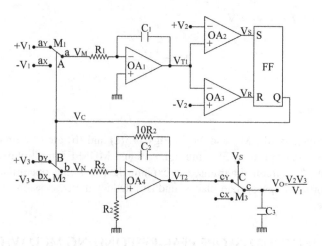

**FIGURE 7.5(B)** Double dual-slope peak-sampling MCD with flip flop

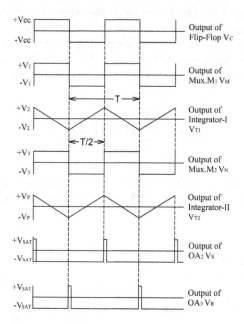

**FIGURE 7.6** Associated waveforms of Fig. 7.5

to HIGH. The multiplexer $M_1$ selects $(+V_1)$ to the integrator I composed by resistor $R_1$, capacitor $C_1$, and op-amp $OA_1$ ('ay' is connected to 'a'). The integrator I output is given as

$$V_{T1} = -\frac{1}{R_1 C_1} \int (+V_1) dt = -\frac{V_1}{R_1 C_1} t \qquad (7.16)$$

## Peak-Responding MCDs with Multiplexers

The output of integrator I is reversing towards negative saturation and when it reaches the value '$(-V_2)$,' the comparator $OA_3$ output becomes HIGH and resets the flip flop so that its output becomes LOW. The multiplexer $M_1$ selects $(-V_1)$ and the sequence repeats to give (i) a triangular waveform $V_{T1}$ of $\pm V_2$ peak-to-peak values with a time period of 'T' at the output of integrator I, (ii) a square waveform $V_C$ at the output of flip flop, and (iii) another square waveform $V_M$ at the output of multiplexer $M_1$. From the waveforms shown in Fig 7.6, equation (7.15) and the fact that at $t = T/2$, $V_{T1} = 2V_2$,

$$2V_2 = \frac{V_1}{R_1 C_1} \frac{T}{2}$$

$$T = \frac{4V_2}{V_1} R_1 C_1 \tag{7.17}$$

The multiplexer $M_2$ connects $+V_3$ during HIGH of the square waveform $V_C$ ('by' is connected to 'b') and $-V_3$ during LOW of the square waveform ('bx' is connected to 'b') $V_C$. Another square waveform $V_N$ with $\pm V_3$ peak-to-peak value is generated at the output of multiplexer $M_2$. This square wave $V_N$ is converted into triangular wave $V_{T2}$ by the integrator II composed by resistor $R_2$, capacitor $C_2$, and op-amp $OA_4$ with $\pm V_P$ as peak-to-peak values of the same time period T. For one transition the integrator II output is given as

$$V_{T2} = -\frac{1}{R_2 C_2} \int (-V_3) dt = \frac{V_3}{R_2 C_2} t \tag{7.18}$$

From the waveforms shown in Fig. 7.6 the equation (7.18), and the fact that at $t = T/2$, $V_{T2} = 2V_p$.

$$2V_P = \frac{V_3}{R_2 C_2} \frac{T}{2}$$

$$V_P = \frac{V_2 V_3}{V_1} \frac{R_1 C_1}{R_2 C_2}$$

Let $R_1 = R_2$ and $C_1 = C_2$

$$V_P = \frac{V_2 V_3}{V_1} \tag{7.19}$$

(i) In Fig. 7.5(a), the peak detector realized by op-amp $OA_5$, diode $D_1$, and capacitor $C_3$ gives a peak value '$V_p$' of the triangular wave $V_{T2}$ and hence $V_O = V_P$.

(ii) In Fig. 7.5(b), the sample and hold circuit realized by multiplexer $M_3$ and capacitor $C_3$ gives the peak value $V_P$ of the triangular wave $V_{T2}$. The short pulse $V_S$ generated at the output of comparator $OA_2$ is acting as a sampling pulse. The sample and hold output is $V_O = V_P$.

From equation (7.19) $V_O = V_P$

$$V_O = \frac{V_2 V_3}{V_1} \qquad (7.20)$$

Design exercise:

1. The multiplexers $M_1$ and $M_2$ in Figs. 7.5(a) and (b) are to be replaced with a transistorized multiplexer, FET multiplexers and MOSFET multiplexers. (i) Draw the circuit diagrams, (ii) explain the working operation, (iii) draw the waveforms at appropriate places, and (iv) deduce the expression for their outputs.
2. In the MCD circuit shown in Fig. 7.5(a) and (b), if the input terminals of multiplexer $M_2$ are reversed and the sampling pulse $V_S$ is replaced by $V_R$, (i) draw the circuit diagrams, (ii) explain the working operation, (iii) draw waveforms at appropriate places, and (iv) deduce the expression for their outputs.

## 7.4 PULSE-WIDTH INTEGRATED PEAK-RESPONDING MCD

The circuit diagrams of pulse-width integrated peak-responding MCDs are shown in Fig. 7.7 and their associated waveforms are shown in Fig. 7.8. Fig. 7.7(a) shows a pulse-width integrated peak-detecting MCD and Fig. 7.7(b) shows a pulse-width integrated peak-sampling MCD. Let's initially assume the comparator $OA_2$ output is LOW, the multiplexer $M_1$ selects 'ax' to 'a' and integrator formed by resistor $R_1$, capacitor $C_1$ and op-amp $OA_1$ integrates ($-V_1$). The integrated output is given as

$$V_{S1} = -\frac{1}{R_1 C_1} \int -V_1 dt = \frac{V_1}{R_1 C_1} t \qquad (7.21)$$

When the output of op-amp $OA_1$ is rising towards positive saturation and it reaches the value '+$V_R$,' the comparator $OA_2$ output will become HIGH, the multiplexer $M_1$ connects 'ay' to 'a,' the capacitor $C_1$ is short circuited and the op-amp $OA_1$ output becomes zero. Now the comparator $OA_2$ output changes to LOW and the cycle therefore repeats to give (i) a sawtooth waveform $V_{S1}$ of peak value $V_R$ and time

# Peak-Responding MCDs with Multiplexers

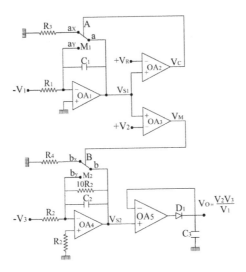

**FIGURE 7.7(A)** Pulse-width integrated peak-detecting MCD

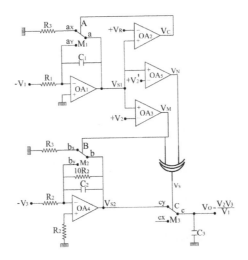

**FIGURE 7.7(B)** Pulse-width integrated peak-sampling MCD

period T at the output of op-amp $OA_1$ and (ii) a short pulse waveform $V_C$ at the output of comparator $OA_2$. From the waveforms shown in Fig. 7.8 and the fact that at $t = T$, $V_{S1} = V_R$ we get

$$V_R = \frac{V_1}{R_1 C_1} T, \quad T = \frac{V_R}{V_1} R_1 C_1 \tag{7.22}$$

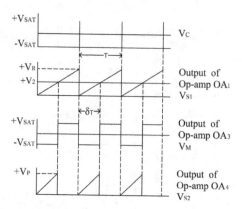

**FIGURE 7.8(A)**  Associated waveforms of Fig. 7.7(a)

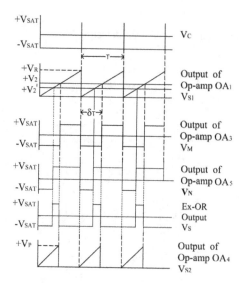

**FIGURE 7.8(B)**  Associated waveforms of Fig. 7.7(b)

The sawtooth waveform $V_{S1}$ is compared with the second input voltage $V_2$ by the comparator $OA_3$. An asymmetrical rectangular wave $V_M$ is generated at the output of comparator $OA_3$. The OFF time of this wave $V_M$ is given as

$$\delta_T = \frac{V_2}{V_R} T \qquad (7.23)$$

The output of comparator $OA_3$ is given as the control input of multiplexer $M_2$. During the ON time of $V_M$, the multiplexer $M_2$ selects 'by' to 'b' and the capacitor $C_2$ is shorted so that zero volts appears at the op-amp $OA_4$ output. During the OFF time of

## Peak-Responding MCDs with Multiplexers

$V_M$, the multiplexer $M_2$ selects 'bx' to 'b,' and another integrator is formed by resistor $R_2$, capacitor $C_2$ and op-amp $OA_3$. This integrator integrates the third input voltage $(-V_3)$ and its output is given as

$$V_{S2} = -\frac{1}{R_2 C_2} \int -V_3 dt = \frac{V_3}{R_2 C_2} t \quad (7.24)$$

A semi-sawtooth wave $V_{S2}$ with peak values of $V_P$ is generated at the output of op-amp $OA_4$. From the waveforms shown in Fig. 7.8, from the equation (7.24) and fact that at $t = \delta_T$, $V_{S2} = V_P$ we get

$$V_P = \frac{V_3}{R_2 C_2} \delta_T \quad (7.25)$$

Equations (7.22) and (7.23) in (7.25) give

$$V_P = \frac{V_2 V_3}{V_1} \frac{R_1 C_1}{R_2 C_2}$$

Let us assume $R_1 C_1 = R_2 C_2$

$$V_P = \frac{V_2 V_3}{V_1}$$

(i) In Fig. 7.7(a), the peak detector realized by diode $D_1$ and capacitor $C_3$ gives this peak value $V_P$ at its output. $V_O = V_P$.
(ii) In Fig. 7.7(b), the peak value $V_P$ is obtained by the sample and hold circuit realized by multiplexer $M_3$ and capacitor $C_3$. The sampling pulse $V_S$ is generated by Ex-OR gate from the signals $V_M$ and $V_N$. $V_N$ is obtained by comparing slightly less than voltage of $V_2$ i.e. $V_2$' with the sawtooth waveform $V_{S1}$. The sampled output is given as $V_O = V_P$.

$$V_O = \frac{V_2 V_3}{V_1} \quad (7.26)$$

### 7.5 MCD USING VOLTAGE TUNABLE ASTABLE MULTIVIBRATOR

The peak-responding MCDs using voltage tunable astable multivibrator are shown in Fig. 7.9 and their associated waveforms are shown in Fig. 7.10. Fig. 7.9(a) shows a peak-detecting MCD and Fig. 7.9(b) shows a peak-sampling MCD. The op-amp $OA_1$ in association with multiplexers $M_1$ and $M_2$ along with the components $R_1$, $R_2$, $R_3$, and $C_1$ produces a square waveform $V_C$. The time period of this square waveform

# Multiplier-Cum-Divider Circuits

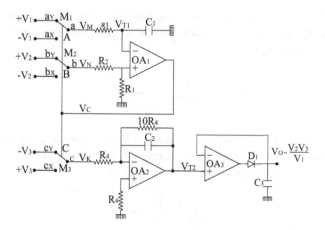

**FIGURE 7.9(A)** Peak-detecting MCD using a voltage tunable astable multivibrator

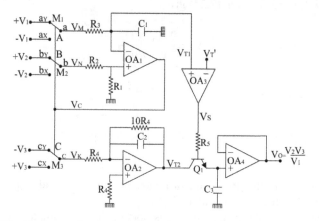

**FIGURE 7.9(B)** Peak-sampling MCD with a voltage tunable astable multivibrator

$V_C$ is proportional to the input voltage $V_2$ and inversely proportional to another input voltage $V_1$

$$T = K \frac{V_2}{V_1} \tag{7.27}$$

where 'K' is constant.

During the LOW value of VC, the multiplexer $M_3$ selects $+V_3$ ('cx' is connected to 'c') and during HIGH value of $V_C$ ($-V_3$) ('cy' is connected to 'c') is connected to the integrator by the multiplexer $M_3$. Another square wave $V_K$ with $\pm V_3$ peak-to-peak value is generated at the multiplexer $M_3$ output. This square wave $V_K$ is converted

# Peak-Responding MCDs with Multiplexers

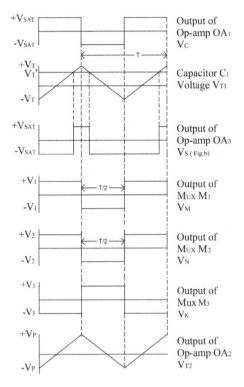

**FIGURE 7.10** Associated waveforms of Fig. 7.9

into triangular wave $V_{T2}$ with a $\pm V_p$ peak-to-peak value by the integrator formed by resistor $R_4$, capacitor $C_2$, and op-amp $OA_2$. For one transition the integrator output is given as

$$V_{T2} = -\frac{1}{R_4 C_2}\int(-V_3)dt = \frac{V_3}{R_4 C_2}t \qquad (7.28)$$

From the waveforms shown in Fig. 7.10 the equation (7.28) and the fact that at $t = T/2$, $V_{T2} = 2V_p$ we get

$$2V_P = \frac{V_3}{R_4 C_2}\frac{T}{2}$$

$$V_P = \frac{V_2 V_3}{V_1}\frac{K}{R_4 C_2}$$

Let us assume $K = R_4C_2$

$$V_P = \frac{V_2 V_3}{V_1} \tag{7.29}$$

(i) In Fig. 7.9(a), the peak detector realized by op-amp $OA_3$, diode $D_1$ and capacitor $C_3$ gives a peak value '$V_P$' of triangular wave $V_{T2}$ and hence $V_O = V_P$.

(ii) In Fig. 7.9(b), the sample and hold circuit realized by transistor $Q_1$ and capacitor $C_3$ gives the peak value $V_P$ of the triangular wave $V_{T2}$. The sampling pulse $V_S$ is generated by comparing the waveform VT1 with a voltage that is slightly less than voltage of VT ($V_T'$). The sample and hold output is $V_O = V_P$.

From equation (7.29)

$$V_O = \frac{V_2 V_3}{V_1} \tag{7.30}$$

# 8 Sawtooth Wave-Referenced MCDs Using Analog Switches

A pulse train whose maximum value is proportional to one voltage ($V_3$) is generated. If the width of this pulse train is made proportional to one voltage ($V_2$) and inversely proportional to another voltage ($V_1$), then the average value of pulse train is proportional to $\dfrac{V_2 V_3}{V_1}$. This is called a time-division multiplier-cum-divider (MCD). There are six types of time-division MCDs.

   (i)   Sawtooth wave-referenced MCD – multiplexing
   (ii)  Triangular wave-referenced MCD – multiplexing
   (iii) MCD using no reference – multiplexing
   (iv)  Sawtooth wave-referenced MCD – switching
   (v)   Triangular wave-referenced MCD – switching
   (vi)  MCD using no reference – switching

The time-division MCD using sawtooth wave-reference – switching – is discussed in section 8.2.

A pulse train whose OFF time proportional to one voltage $V_2$ and inversely proportional to another voltage $V_1$ is generated. The third input voltage $V_3$ is integrated during this OFF time. The peak value of this integrated output is proportional to $\dfrac{V_2 V_3}{V_1}$. This is called time-division peak-detecting MCD and is discussed in section 8.3.

## 8.1 SAWTOOTH WAVE GENERATORS

Two circuits for the generation of a sawtooth wave using analog switches are given in Figs. 8.1(a) and (b) and their associated waveforms are shown in Fig. 8.2. A sawtooth wave $V_{S1}$ of peak value $V_R$ and time period T is generated by these circuits.

In Fig. 8.1(a), initially op-amp $OA_2$ output is LOW, the switch $S_1$ is open and the integrator formed by resistor $R_1$, capacitor $C_1$ and op-amp $OA_1$ integrates ($-V_R$) and its output is given as

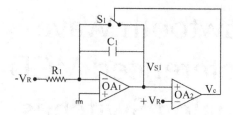

**FIGURE 8.1(A)** Sawtooth wave generator – I

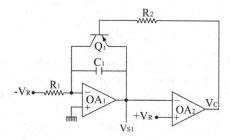

**FIGURE 8.1(B)** Sawtooth wave generator – II

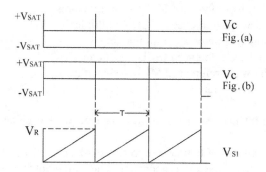

**FIGURE 8.2** Associated waveforms of Fig. 8.1

$$V_{S1} = -\frac{1}{R_1 C_1} \int -V_R \, dt$$

$$V_{S1} = \frac{V_R}{R_1 C_1} t \tag{8.1}$$

A positive-going ramp is generated at the output of op-amp $OA_1$ and when it reaches the value of reference voltage $+V_R$ the comparator $OA_2$ output becomes HIGH. The switch $S_1$ closes and shorts capacitor $C_1$ and hence the integrator output becomes zero. Then comparator output is LOW and the sequence therefore repeats to give a perfect sawtooth wave $V_{S1}$ of peak value $V_R$ at the op-amp $OA_1$ output as shown in Fig. 8.2. From equation (8.1), Fig. 8.2 and fact that at t= T, $V_{S1} = V_R$ we get

$$V_R = \frac{V_R}{R_1 C_1} T$$

$$T = R_1 C_1 \tag{8.2}$$

In Fig. 8.1(b), initially the op-amp $OA_2$ output is HIGH. The transistor $Q_1$ is OFF and the integrator formed by resister $R_1$, capacitor $C_1$ and op-amp $OA_1$ integrates $(-V_R)$ and its output is given as

$$V_{S1} = -\frac{1}{R_1 C_1} \int -V_R dt$$

$$V_{S1} = \frac{V_R}{R_1 C_1} t \tag{8.3}$$

A positive-going ramp is generated at the output of op-amp $OA_1$ and when it reaches the value of reference voltage $(+V_R)$ the comparator $OA_2$ output becomes LOW. The transistor $Q_1$ is ON and shorts capacitor $C_1$ and hence the integrator output becomes zero. Then the comparator output is HIGH and the sequence therefore repeats to give a perfect sawtooth wave $V_{S1}$ of peak value $V_R$ at op-amp $OA_1$ output and a short pulse waveform $V_C$ at the output of op-amp $OA_2$ as shown in Fig. 8.2, from equation (8.3), Fig. 8.2 and fact that at t = T, $V_{S1} = V_R$ we get

$$V_R = \frac{V_R}{R_1 C_1} T$$

$$T = R_1 C_1 \tag{8.4}$$

## 8.2 DOUBLE SWITCHING AND AVERAGING TIME-DIVISION MCD

The circuit diagrams of double switching and averaging time-division MCDs are shown in Fig. 8.3 and their associated waveforms are shown in Fig. 8.4. Fig. 8.3(a) shows a series switching MCD and Fig. 8.3(b) shows a shunt-switching MCD. As discussed in section 8.1, a sawtooth wave $V_{S1}$ of peak value $V_R$ time period T is generated by op-amps $OA_1$, $OA_2$, and switch $S_1$.

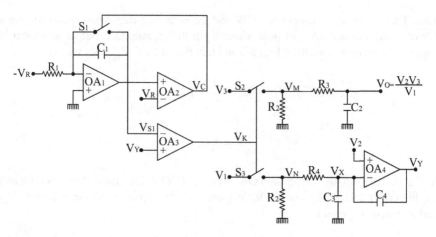

**FIGURE 8.3(A)** Double series switching – averaging time-division MCD

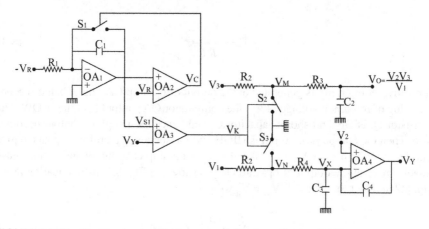

**FIGURE 8.3(B)** Double shunt switching – averaging time-division MCD

The comparator $OA_3$ compares the sawtooth wave with the voltage $V_Y$ and produces a first rectangular waveform $V_K$. The ON time [Fig. 8.3(a)] or OFF time [Fig. 8.3(b)] $\delta_T$ of $V_K$ is given as

$$\delta_T = \frac{V_Y}{V_R} T \tag{8.5}$$

The rectangular pulse VK controls the switches S2 and S3.

(i) In Fig. 8.3(a), when $V_K$ is HIGH, the switch $S_2$ is closed and the third input voltage $V_3$ is connected to the $R_3 C_2$ low pass filter, the switch $S_3$ is closed and

## Sawtooth Wave-Referenced MCD

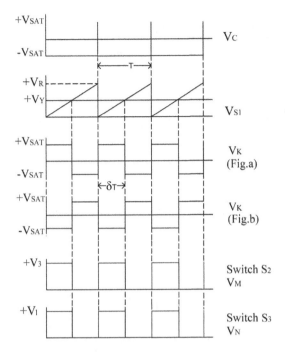

**FIGURE 8.4** Associated waveforms of Fig. 8.3

the first input voltage $V_1$ is connected to the $R_4C_3$ low pass filter. When $V_K$ is LOW, the switch $S_2$ is open and zero volts exists on the $R_3C_2$ low pass filter, the switch $S_3$ is open and zero volts exists on the $R_4C_3$ low pass filter.

(i) In Fig. 8.3(b), when $V_K$ is HIGH, the switch $S_2$ is closed and zero volts exists on the $R_3C_2$ low pass filter, the switch $S_3$ is closed and zero volts exists on the $R_4C_3$ low pass filter. When $V_K$ is LOW, the switch $S_2$ is open and the third input voltage $V_3$ is connected to the $R_3C_2$ low pass filter, the switch $S_3$ is open and the first input voltage $V_1$ is connected to the $R_4C_3$ low pass filter.

The second rectangular pulse $V_N$ with a maximum value of $V_1$ is generated at the switch $S_3$ output. The $R_4C_3$ low pass filter gives the average value of this pulse train $V_N$ and is given as

$$V_X = \frac{1}{T}\int_0^{\delta_T} V_1 dt = \frac{V_1}{T}\delta_T$$

$$V_X = \frac{V_1 V_Y}{V_R} \qquad (8.6)$$

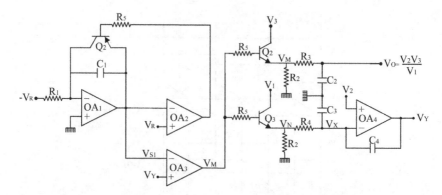

**FIGURE 8.5(A)** Double series switching – averaging time-division MCD

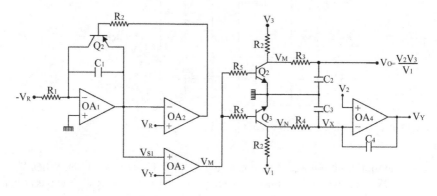

**FIGURE 8.5(B)** Double shunt switching – averaging time-division MCD

The op-amp OA4 is configured in a negative closed loop feedback and a positive DC voltage is ensured in the feedback loop. Hence its inverting terminal voltage must be equal to its non-inverting terminal voltage.

$$V_2 = V_X \tag{8.7}$$

From equations (8.6) and (8.7)

$$V_Y = \frac{V_2 V_R}{V_1} \tag{8.8}$$

The third rectangular pulse $V_M$ with the maximum value of $V_3$ is generated at the switch $S_2$ output. The $R_3 C_2$ low pass filter gives the average value of this pulse train $V_M$ and is given as

$$V_O = \frac{1}{T} \int_0^{\delta_T} V_3 dt = \frac{V_3}{T} \delta_T \tag{8.9}$$

Equations (8.5) and (8.8) in (8.9) give

$$V_O = \frac{V_2 V_3}{V_1} \quad (8.10)$$

Design exercise:

1. The switches $S_2$ and $S_3$ in Figs. 8.3(a) and (b) are replaced with transistor switches and shown in Figs. 8.5(a) and (b). (i) Explain the working operation of the MCDs shown in Figs. 8.5(a) and (b), (ii) draw waveforms in appropriate places, and (iii) deduce the expression for their outputs.
2. The switches $S_2$ and $S_3$ in Figs. 8.3(a) and (b) are to be replaced with FET switches and MOSFET switches. In each, (i) draw the circuit diagrams, (ii) explain their working operations, (iii) draw waveforms at appropriate places, and (iv) deduce the expression for their output voltages.
3. The circuits shown in Figs. 8.3(a) and (b) use the sawtooth wave of Fig. 8.1(a). Replace this sawtooth wave generator with the other sawtooth generator shown in Fig. 8.1(b). (i) Draw the circuit diagrams, (ii) explain their working operations, (iii) draw waveforms at appropriate places, and (iv) deduce the expression for their output voltages.

## 8.3 TIME-DIVISION SINGLE-SLOPE PEAK-DETECTING MCD

The circuit diagram of a time-division single-slope peak-detecting MCD is shown in Fig. 8.6 and its associated waveforms in Fig. 8.7. A sawtooth wave $V_{S1}$ is generated by resistor $R_1$, capacitor $C_1$, and op-amp $OA_1$.

The comparator $OA_2$ compares the sawtooth wave $V_{S1}$ with the voltage $V_Y$ and produces a rectangular waveform $V_M$. The OFF time $\delta_T$ of $V_M$ is given as

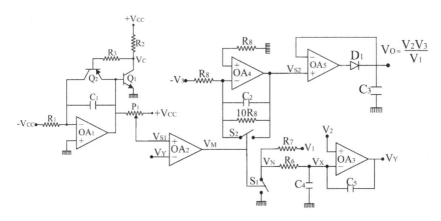

**FIGURE 8.6** Time-division single-slope peak-detecting MCD

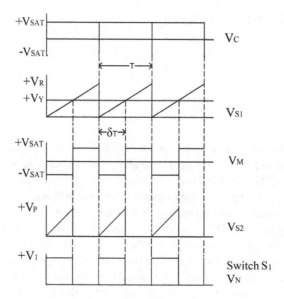

**FIGURE 8.7** Associated waveforms of Fig. 8.6

$$\delta_T = \frac{V_Y}{V_R} T \tag{8.11}$$

The rectangular pulse $V_M$ controls the switch $S_1$. When $V_M$ is LOW, switch $S_1$ is open, and input voltage $V_1$ is connected to the $R_6C_4$ low pass filter. When $V_M$ is HIGH, switch $S_1$ is closed, and zero volts are connected to the $R_6C_4$ low pass filter. Another rectangular pulse $V_N$ with maximum value of $V_1$ is generated at the switch $S_1$ output. The $R_6C_4$ low pass filter gives the average value of this pulse train $V_N$ and is given as

$$V_X = \frac{1}{T}\int_0^{\delta_T} V_1 dt = \frac{V_1}{T}\delta_T$$

$$V_X = \frac{V_1 V_Y}{V_R} \tag{8.12}$$

The op-amp $OA_3$ is configured in a negative closed loop feedback and a positive DC voltage is ensured in the feedback loop. Hence its inverting terminal voltage must be equal to its non-inverting terminal voltage.

$$V_2 = V_X \tag{8.13}$$

# Sawtooth Wave-Referenced MCD

From equations (8.12) and (8.13)

$$V_Y = \frac{V_2}{V_1} V_R \qquad (8.14)$$

The rectangular pulse $V_M$ also controls switch $S_2$. During the HIGH of $V_M$, switch $S_2$ is closed and hence the capacitor $C_2$ is short circuited so that the op-amp $OA_4$ output is zero volts. During the LOW of $V_M$, switch $S_2$ is open and the integrator formed by resistor $R_8$, capacitor $C_2$, op-amp $OA_4$ integrates its input voltage $(-V_3)$ and its output is given as

$$V_{S2} = -\frac{1}{R_8 C_2}\int -V_3 dt = \frac{V_3}{R_8 C_2} t \qquad (8.15)$$

A semi-sawtooth waveform $V_{S2}$ with peak value $V_P$ is generated at the output of op-amp $OA_4$. From the waveforms shown in Fig. 8.7 and from equation (8.15) the fact that at $t = \delta_T$, $V_{S2} = V_P$ we get

$$V_P = \frac{V_3}{R_2 C_2} \delta_T \qquad (8.16)$$

Equation (8.11) in (8.16) gives

$$V_P = \frac{V_3}{R_2 C_2} \frac{V_Y}{V_R} T = \frac{V_3}{R_2 C_2} \frac{V_2}{V_1} T$$

Let us assume $T = R_2 C_2$

$$V_P = \frac{V_2 V_3}{V_1} \qquad (8.17)$$

The peak detector realized by op-amp $OA_5$, diode $D_1$ and capacitor $C_3$ gives this peak value $V_P$ at its output $V_O$. $V_O = V_P$. Hence the output will be

$$V_O = \frac{V_2 V_3}{V_1} \qquad (8.18)$$

Design exercise:

1. The switches $S_1$ and $S_2$ in Fig. 8.6 are replaced with transistor switches and shown in Fig. 8.8. (i) Explain the working operation of the MCDs shown in

# Multiplier-Cum-Divider Circuits

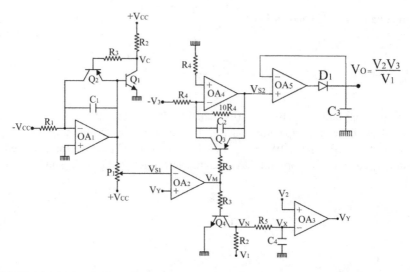

**FIGURE 8.8** Time-division single-slope peak-detecting MCD

Fig. 8.8, (ii) draw waveforms in appropriate places, and (iii) deduce expression for their outputs.

2. The switches $S_1$ and $S_2$ in Fig. 8.6 are to be replaced with FET switches and MOSFET switches. In each, (i) draw the circuit diagrams, (ii) explain their working operations, (iii) draw waveforms at appropriate places, and (iv) deduce the expression for their output voltages.
3. The circuits shown in Fig. 8.6 uses the sawtooth wave of Fig. 5.1(d). Replace this sawtooth wave generator with the other sawtooth generators shown in Figs. 8.1(a) and (b). In each, (i) draw the circuit diagrams, (ii) explain their working operations, (iii) draw waveforms at appropriate places, and (iv) deduce expression for their output voltages.

## 8.4 TIME-DIVISION MULTIPLY–DIVIDE MCD

The circuit diagrams of sawtooth wave-based time-division multiply–divide MCDs are shown in Fig. 8.9 and their associated waveforms are shown in Fig. 8.10. Fig. 8.9(a) shows a series switching MCD and Fig. 8.9(b) shows a shunt-switching MCD. A sawtooth wave $V_{S1}$ of period T is generated around the operational amplifier $OA_1$. The comparator $OA_2$ compares the sawtooth wave with an input voltage $V_3$ and produces a rectangular waveform $V_M$. The ON time $\delta_{T1}$ of $V_M$ [Fig. 8.9(a)] or OFF time $\delta_{T1}$ of $V_M$ [Fig. 8.9(b)] is given as

$$\delta_{T1} = \frac{V_3}{V_R} T \qquad (8.19)$$

The rectangular pulse $V_M$ controls the first switch $S_1$.

## Sawtooth Wave-Referenced MCD

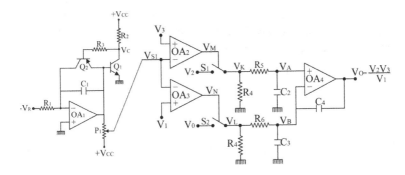

**FIGURE 8.9(A)**  Series switching time-division multiply–divide MCD

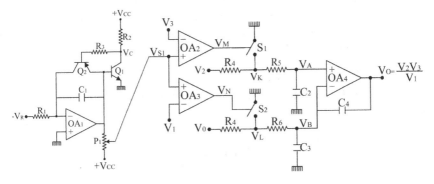

**FIGURE 8.9(B)**  Shunt switching time-division multiply–divide MCD

(i) In Fig. 8.9(a), When $V_M$ is HIGH, switch $S_1$ is closed, another input voltage $V_2$ is connected to the $R_5C_2$ low pass filter. When $V_M$ is LOW, switch $S_1$ is open, zero volts is connected to the $R_5C_2$ low pass filter.

(ii) In Fig. 8.9(b), When $V_M$ is HIGH, the switch $S_1$ is closed, and zero volts are connected to the $R_5C_2$ low pass filter. When $V_M$ is LOW, $S_1$ is open, and another input voltage $V_2$ is connected to $R_5C_2$ low pass filter.

Another rectangular pulse $V_K$ with the maximum value of $V_2$ is generated at the switch $S_1$ output. The $R_5C_2$ low pass filter gives the average value of this pulse train $V_K$ and is given as

$$V_A = \frac{1}{T}\int_0^{\delta_{T1}} V_2 \, dt = \frac{V_2}{T}\delta_{T1}$$

$$V_A = \frac{V_2 V_3}{V_R} \tag{8.20}$$

The comparator OA$_3$ compares the sawtooth wave V$_{S1}$ with the first input voltage V$_1$ and produces a rectangular waveform V$_N$. The ON time $\delta_{T2}$ of V$_N$ [Fig. 8.9(a)] or OFF time $\delta_{T2}$ of V$_N$ [Fig. 8.9(b)] is given as

$$\delta_{T2} = \frac{V_1}{V_R}T \qquad (8.21)$$

The rectangular pulse V$_N$ controls the second switch S$_2$.

(i) In Fig 8.9(a), when V$_N$ is HIGH, the output voltage V$_O$ is connected to the R$_6$C$_3$ low pass filter (switch S$_2$ is closed). When V$_N$ is LOW zero volts are connected to the R$_6$C$_3$ low pass filter (switch S$_2$ is open).
(ii) In Fig 8.9(b), when V$_N$ is HIGH, zero volts is connected to the R$_6$C$_3$ low pass filter (switch S$_2$ is closed). When V$_N$ is LOW, output voltage V$_O$ is connected to R$_6$C$_3$ low pass filter (switch S$_2$ is open).

Another rectangular pulse V$_L$ with maximum value of V$_O$ is generated at the switch S$_2$ output. The R$_6$C$_3$ low pass filter gives average value of this pulse train V$_L$ and is given as

$$V_B = \frac{1}{T}\int_0^{\delta_{T2}} V_O\,dt = \frac{V_O}{T}\delta_{T2}$$

$$V_B = \frac{V_1 V_O}{V_R} \qquad (8.22)$$

The op-amp OA$_4$ is configured in a negative closed loop feedback and a positive DC voltage is ensured in the feedback loop. Hence its inverting terminal voltage must be equal to its non-inverting terminal voltage, i.e.

$$V_A = V_B \qquad (8.23)$$

From equations (8.20) and (8.22)

$$V_O = \frac{V_2 V_3}{V_1} \qquad (8.24)$$

Design exercise:

1. The switches S$_1$ and S$_2$ in Figs 8.9(a) and (b) are replaced with transistor switches and shown in Figs. 8.11(a) and (b). (i) Explain the working operation of the MCDs shown in Figs. 8.11(a) and (b), (ii) draw waveforms in appropriate places, and (iii) deduce the expression for their outputs.

# Sawtooth Wave-Referenced MCD

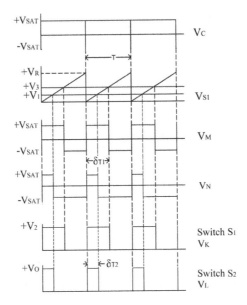

**FIGURE 8.10(A)** Associated waveforms of Fig. 8.9(a)

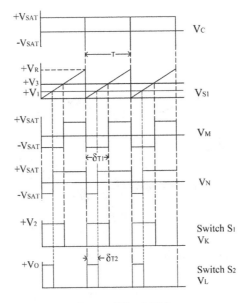

**FIGURE 8.10(B)** Associated waveforms of Fig. 8.9(b)

2. The switches $S_1$ and $S_2$ in Figs. 8.9(a) and (b) are to be replaced with FET switches and MOSFET switches. In each, (i) draw the circuit diagrams, (ii) explain their working operations, (iii) draw waveforms at appropriate places, and (iv) deduce the expression for their output voltages.

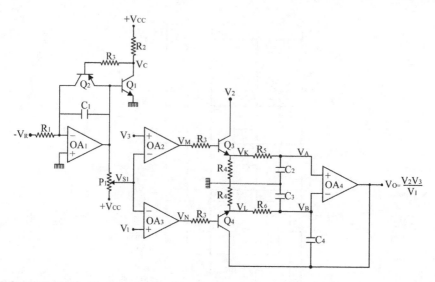

**FIGURE 8.11(A)** Series switching time-division multiply–divide MCD

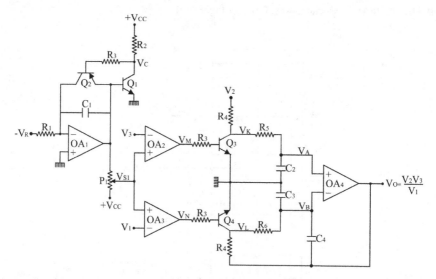

**FIGURE 8.11(B)** Shunt switching time-division multiply–divide MCD

3. The circuits shown in Figs. 8.9(a) and (b) uses sawtooth wave of Fig. 5.1(d). Replace this sawtooth wave generator with the other sawtooth generator shown in Figs. 8.1(a) and (b). (i) Draw the circuit diagrams, (ii) explain their working operations, (iii) draw waveforms at appropriate places, and (iv) deduce the expression for their output voltages

# Sawtooth Wave-Referenced MCD

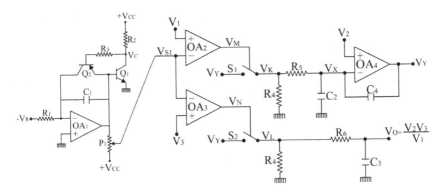

**FIGURE 8.12(A)**  Series switching time-division divide–multiply MCD

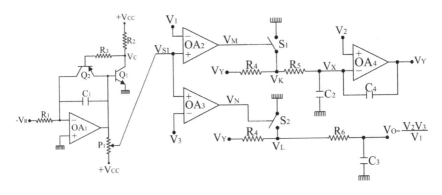

**FIGURE 8.12(B)**  Shunt switching time-division divide–multiply MCD

## 8.5 TIME-DIVISION DIVIDE–MULTIPLY MCD

The circuit diagrams of sawtooth wave-based time-division divide–multiply MCDs are shown in Fig. 8.12 and their associated waveforms are shown in Fig. 8.13. Fig. 8.12(a) shows a series-switching MCD and Fig. 8.12(b) shows a shunt-switching MCD. A sawtooth wave $V_{S1}$ of $V_R$ peak value and time period T is generated by around the op-amp $OA_1$. The comparator $OA_2$ compares the sawtooth wave with an input voltage $V_1$ and produces a rectangular waveform $V_M$. The ON time $\delta_{T1}$ of $V_M$ [Fig. 8.12(a)] or OFF time $\delta_{T1}$ of $V_M$ [Fig. 8.12(b)] is given as

$$\delta_{T1} = \frac{V_1}{V_R} T \qquad (8.25)$$

The rectangular pulse $V_M$ controls the first switch $S_1$.

(i) In Fig. 8.12(a), when $V_M$ is HIGH, the switch $S_1$ is closed, and the voltage $V_Y$ is connected to the $R_5C_2$ low pass filter. When $V_M$ is LOW, the switch $S_1$ is open, and zero volts are connected to the $R_5C_2$ low pass filter.

(ii) In Fig. 8.12(b), when $V_M$ is HIGH, the switch $S_1$ is closed, and zero volts are connected to the $R_5C_2$ low pass filter. When $V_M$ is LOW, the switch $S_1$ is open, and input voltage $V_Y$ is connected to the $R_5C_2$ low pass filter.

Another rectangular pulse $V_K$ with the maximum value of $V_Y$ is generated at the switch $S_1$ output. The $R_5C_2$ low pass filter gives the average value of this pulse train $V_N$ and is given as

$$V_X = \frac{1}{T}\int_0^{\delta_{T1}} V_Y dt = \frac{V_Y}{T}\delta_{T1}$$

$$V_X = \frac{V_1 V_Y}{V_R} \tag{8.26}$$

The op-amp $OA_4$ is configured in a negative closed loop feedback and a positive DC voltage is ensured in the feedback loop. Hence its inverting terminal voltage must be equal to its non-inverting terminal voltage, i.e.

$$V_X = V_2 \tag{8.27}$$

From equations (8.26) and (8.27)

$$V_Y = \frac{V_2 V_R}{V_1} \tag{8.28}$$

The comparator $OA_3$ compares the sawtooth wave $V_{S1}$ with the third input voltage $V_3$ and produces a rectangular waveform $V_N$. The ON time $\delta_{T2}$ of $V_N$ [Fig. 8.12(a)] or OFF time $\delta_{T2}$ of $V_N$ [Fig. 8.12(b)] is given as

$$\delta_{T2} = \frac{V_3}{V_R} T \tag{8.29}$$

The rectangular pulse $V_N$ controls the second switch $S_2$.

(i) In Fig. 8.12(a), when $V_N$ is HIGH, the switch $S_2$ is closed, and the voltage $V_Y$ is connected to the $R_6C_3$ low pass filter. When $V_N$ is LOW, the switch $S_2$ is open, and zero volts is connected to the $R_6C_3$ low pass filter.
(ii) In Fig. 8.12(b), when $V_N$ is HIGH, the switch $S_2$ is closed, and zero volts are connected to the $R_6C_3$ low pass filter. When $V_N$ is LOW, the switch $S_2$ is open, and the voltage $V_Y$ is connected to the $R_6C_3$ low pass filter.

# Sawtooth Wave-Referenced MCD

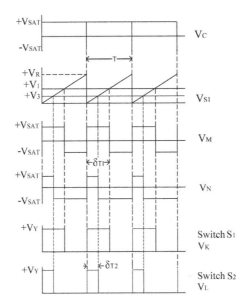

**FIGURE 8.13(A)** Associated waveforms of Fig. 8.12(a)

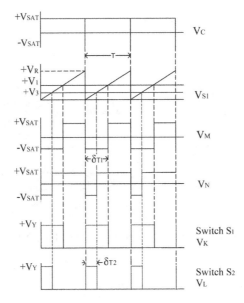

**FIGURE 8.13(B)** Associated waveforms of Fig. 8.12(b)

112                                                         Multiplier-Cum-Divider Circuits

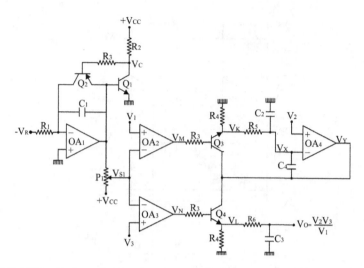

**FIGURE 8.14(A)**  Series switching time-division divide–multiply MCD

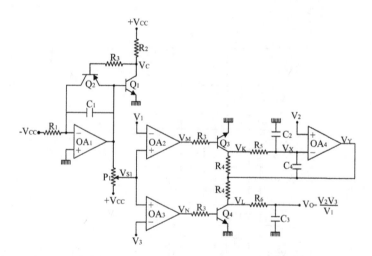

**FIGURE 8.14(B)**  Shunt switching time-division divide–multiply MCD

# Sawtooth Wave-Referenced MCD

Another rectangular pulse $V_L$ with the maximum value of $V_Y$ is generated at the switch $S_2$ output. The $R_6C_3$ low pass filter gives the average value of this pulse train $V_L$ and is given as

$$V_O = \frac{1}{T}\int_0^{\delta_{T2}} V_Y dt = \frac{V_Y}{T}\delta_{T2} \tag{8.30}$$

$$V_O = \frac{V_3 V_Y}{V_R} \tag{8.31}$$

Equation (8.28) in (8.31) gives

$$V_O = \frac{V_2 V_3}{V_1} \tag{8.32}$$

Design exercise:

1. The switches $S_1$ and $S_2$ in Figs 8.12(a) and (b) are replaced with transistor switches and shown in Figs. 8.14(a) and (b). (i) Explain the working operation of the MCDs shown in Figs. 8.14(a) and (b), (ii) draw waveforms in appropriate places, and (iii) deduce the expression for their outputs.
2. The switches $S_1$ and $S_2$ in Figs. 8.12(a) and (b) are to be replaced with FET switches and MOSFET switches. In each, (i) draw the circuit diagrams, (ii) explain their working operations, (iii) draw waveforms at appropriate places, and (iv) deduce the expression for their output voltages.
3. The circuits shown in Figs. 8.12(a) and (b) use the sawtooth wave of Fig. 5.1(d). Replace this sawtooth wave generator with the other sawtooth generators shown in Figs. 8.1(a) and (b). In each, (i) draw the circuit diagrams, (ii) explain their working operations, (iii) draw waveforms at appropriate places, and (iv) deduce the expression for their output voltages.

# 9 Triangular Wave-Referenced MCD with Analog Switches

A pulse train whose maximum value is proportional to one voltage ($V_3$) is generated. If the width of this pulse train is made proportional to one voltage ($V_2$) and inversely proportional to another voltage ($V_1$), then the average value of pulse train is proportional to $\frac{V_2 V_3}{V_1}$. This is called a time-division multiplier-cum-divider (MCD).

There are six types of time-division MCDs.

(i) Sawtooth wave-referenced MCD – multiplexing
(ii) Triangular wave-referenced MCD – multiplexing
(iii) MCD using no reference – multiplexing
(iv) Sawtooth wave-referenced MCD – switching
(v) Triangular wave-referenced MCD – switching
(vi) MCD using no reference – switching

The time-division MCD using triangular wave as reference – switching – is discussed in Section 9.2.

## 9.1 TIME-DIVISION MCD

The circuit diagrams of triangular wave-referenced time-division MCDs are shown in Fig. 9.1 and their associated waveforms are shown in Fig. 9.2. Fig. 9.1(a) shows a series-switching MCD and Fig. 9.1(b) shows a shunt-switching MCD. As discussed

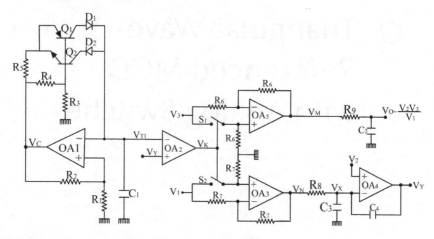

**FIGURE 9.1(A)**  Series switching time-division MCD

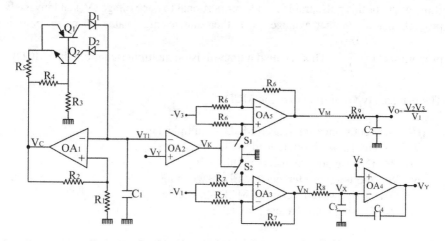

**FIGURE 9.1(B)**  Shunt switching time-division MCD

in Section 6.1, the output of op-amp $OA_1$ is a triangular wave $V_{T1}$ with $\pm V_T$ peak values and time period of T.

The comparator $OA_2$ compares this triangular wave $V_{T1}$ with the voltage $V_Y$ and produces an asymmetrical rectangular wave $V_K$. From Fig 9.2, it is observed that

$$T_1 = \frac{V_T - V_Y}{2V_T}T, \quad T_2 = \frac{V_T + V_Y}{2V_T}T, \quad T = T_1 + T_2 \tag{9.1}$$

The rectangular wave $V_K$ controls switches $S_1$ and $S_2$. During the ON time $T_2$ of this rectangular wave $V_K$

# Triangular Wave-Referenced MCD

(i) In Fig. 9.1(a), the switch $S_1$ is closed, the op-amp $OA_5$ along with resistors $R_6$ will work as a non-inverting amplifier and $+V_3$ will exist on its output ($V_M = +V_3$). The switch $S_2$ is closed, the op-amp $OA_3$ along with resistor $R_7$ will work as a non-inverting amplifier and $+V_1$ will exist on its output ($V_N = +V_1$).

(ii) In Fig. 9.1(b), the switch $S_1$ is closed, the op-amp $OA_5$ along with resistors $R_6$ will work as an inverting amplifier and $+V_3$ will exist on its output ($V_M = +V_3$). The switch $S_2$ is closed, the op-amp $OA_3$ along with resistor $R_7$ will work as an inverting amplifier and $+V_1$ will exist on its output ($V_N = +V_1$).

During OFF time $T_1$ of this rectangular wave $V_K$

(i) In Fig. 9.1(a), the switch $S_1$ is opened, the op-amp $OA_5$ along with resistors $R_6$ will work an inverting amplifier and $-V_3$ will exist on its output ($V_M = -V_3$). The switch $S_2$ is opened, the op-amp $OA_3$ along with resistor $R_7$ will work as an inverting amplifier and $-V_1$ will exist on its output ($V_N = -V_1$).

(ii) In Fig. 9.1(b), the switch $S_1$ is opened, the op-amp $OA_5$ along with resistors $R_6$ will work as a non-inverting amplifier and $-V_3$ will exist on its output ($V_M = -V_3$). The switch $S_2$ is closed, the op-amp $OA_3$ along with resistors $R_7$ will work as non-inverting amplifier and $-V_1$ will exists on its output ($V_N = -V_1$)

Two asymmetrical rectangular waveforms (i) $V_M$ with $\pm V_3$ peak-to-peak values at the output of op-amp $OA_5$ and (ii) $V_N$ with $\pm V_1$ peak-to-peak values at the output of op-amp $OA_3$, are generated. The $R_8 C_3$ low pass filter gives the average value of $V_N$ and is given as

$$V_X = \frac{1}{T}\left[\int_0^{T_2} V_1\, dt + \int_{T_2}^{T_1+T_2}(-V_1)\, dt\right] = \frac{V_1}{T}[T_2 - T_1]$$

$$V_X = \frac{V_1 V_Y}{V_T} \qquad (9.2)$$

The op-amp $OA_4$ is configured in a negative closed loop feedback and a positive DC voltage is ensured in the feedback loop. Hence its inverting terminal voltage must equal to its non-inverting terminal voltage, i.e.

$$V_X = V_2 \qquad (9.3)$$

From equations (9.2) and (9.3)

$$V_Y = \frac{V_2 V_T}{V_1} \qquad (9.4)$$

The $R_9C_2$ low pass filter gives the average value of the rectangular waveform $V_M$ and is given as

$$V_O = \frac{1}{T}\left[\int_0^{T_2} V_3\, dt + \int_{T_2}^{T_1+T_2}(-V_3)\, dt\right] = \frac{V_3}{T}[T_2 - T_1]$$

$$V_O = \frac{V_3 V_Y}{V_T} \tag{9.5}$$

Equation (9.4) in (9.5) gives

$$V_O = \frac{V_2 V_3}{V_1} \tag{9.6}$$

Design exercise:

1. In the MCD circuits shown in Figs. 9.1(a),(b), the triangular wave generator shown in Fig. 6.1(a) is used. Replace this triangular wave generator with the other triangular wave generator shown in Fig. 6.1(b). (i) Draw circuit diagrams, (ii) draw waveforms at appropriate places, (iii) explain the working operation, and (iv) deduce the expression for their output voltages.
2. The switches $S_1$ and $S_2$ in Figs. 9.1(a) and (b) can be replaced with transistor switches and shown in Figs. 9.3(a) and (b). (i) Draw waveforms at appropriate places, (ii) explain the working operation, and (iii) deduce the expression for their output voltages.

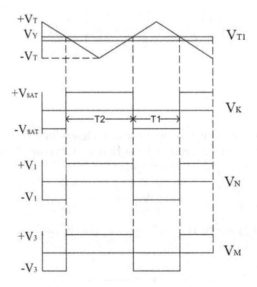

**FIGURE 9.2** Associated waveforms of Fig. 9.1

3. Replace the switches $S_1$ and $S_2$ in Figs. 9.1(a) and (b) with FET switches and MOSFET switches. In each, (i) draw circuit diagrams, (ii) draw waveforms at appropriate places, (iii) explain the working operation, and (iv) deduce the expression for their output voltages.

## 9.2 DIVIDE–MULTIPLY TIME-DIVISION MCD

The circuit diagrams of triangular wave-based divide–multiply time-division MCDs are shown in Fig. 9.4 and their associated waveforms are shown in Fig. 9.5. As discussed in Section 6.1, a triangular wave $V_{T1}$ with a $\pm V_T$ peak-to-peak value is generated by resistors $R_1$ to $R_5$, capacitor $C_1$, the op-amp $OA_1$, and transistors $Q_1$ and $Q_2$.

The first input voltage $V_1$ is compared with the generated triangular wave $V_{T1}$ by the comparator on $OA_2$. An asymmetrical rectangular waveform $V_N$ is generated at the comparator $OA_2$ output. From the waveforms shown in Fig. 9.5, it is observed that

$$T_1 = \frac{V_T - V_1}{2V_T}T \quad T_2 = \frac{V_T + V_1}{2V_T}T \quad T = T_1 + T_2 \quad (9.7)$$

This rectangular wave $V_N$ is given as control input to the switch $S_1$.
During the ON time $T_2$ of this rectangular wave $V_N$

(i) In Fig. 9.4(a), the switch $S_1$ is closed, the op-amp $OA_5$ along with resistors $R_6$ will work as a non-inverting amplifier and $+V_Y$ will exist on its output ($V_K = +V_Y$).
(ii) In Fig. 9.4(b), the switch $S_1$ is closed, the op-amp $OA_5$ along with resistors $R_6$ will work as an inverting amplifier and $+V_Y$ will exist on its output ($V_K = +V_Y$).

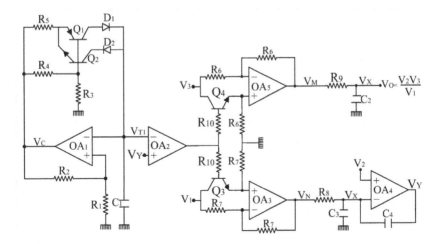

**FIGURE 9.3(A)** Series-switching time-division MCD

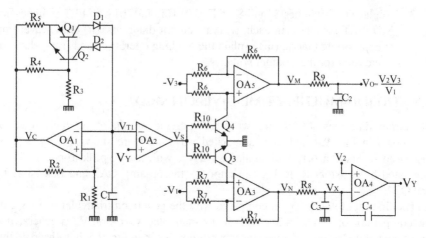

**FIGURE 9.3(B)**  Shunt-switching time-division MCD

During the OFF time $T_3$ of this rectangular wave $V_N$

(iii) In Fig. 9.4(a), the switch $S_1$ is opened, the op-amp $OA_5$ along with resistors $R_6$ will work as an inverting amplifier and $-V_Y$ will exist on its output ($V_K = -V_Y$).

(iv) In Fig. 9.4(b), the switch $S_1$ is opened, the op-amp $OA_5$ along with resistors $R_6$ will work as a non-inverting amplifier and $-V_Y$ will exist on its output ($V_K = -V_Y$).

Another rectangular asymmetrical wave $V_K$ with peak-to-peak values of $\pm V_Y$ is generated at the output of op-amp $OA_5$. The $R_7 C_3$ low pass filter gives the average value of the pulse train $V_K$ and is given as

$$V_X = \frac{1}{T}\left[\int_0^{T_2} V_Y \, dt + \int_{T_2}^{T_1+T_2} (-V_Y) \, dt\right] = \frac{V_Y}{T}(T_2 - T_1)$$

$$V_X = \frac{V_1 V_Y}{V_T} \tag{9.8}$$

The op-amp $OA_4$ is configured in a negative closed loop feedback and a positive DC voltage is ensured in the feedback loop. Hence its inverting terminal voltage must be equal to its non-inverting terminal voltage, i.e.

$$V_2 = V_X \tag{9.9}$$

From equations (9.8) and (9.9)

$$V_Y = \frac{V_2 V_T}{V_1} \tag{9.10}$$

# Triangular Wave-Referenced MCD

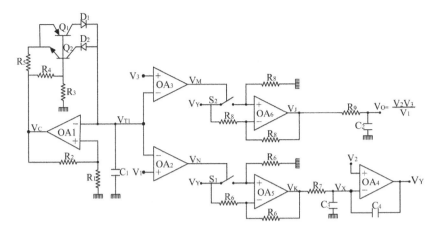

**FIGURE 9.4(A)** Series-switching divide–multiply time-division MCD

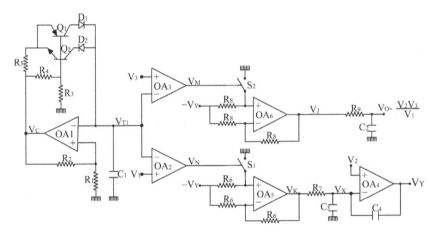

**FIGURE 9.4(B)** Shunt-switching divide–multiply time-division MCD

The third input voltage $V_3$ is compared with the generated triangular wave $V_{T1}$ by the comparator on $OA_3$. An asymmetrical rectangular waveform $V_M$ is generated at the comparator $OA_3$ output. From the waveforms shown in Fig. 9.5, it is observed that

$$T_3 = \frac{V_T - V_3}{2V_T} T$$

$$T_4 = \frac{V_T + V_3}{2V_T} T$$

$$T = T_3 + T_4 \tag{9.11}$$

This rectangular wave $V_M$ is given as the control input to the switch $S_2$.

# Multiplier-Cum-Divider Circuits

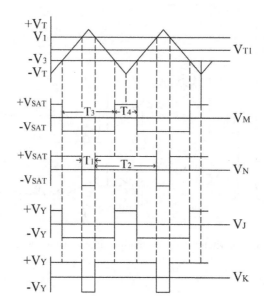

**FIGURE 9.5** Associated waveforms of Fig. 9.4

During the ON time $T_4$ of this rectangular wave $V_M$

(i) In Fig. 9.4(a), the switch $S_2$ is closed, the op-amp $OA_6$ along with resistors $R_8$ will work as a non-inverting amplifier and $+V_Y$ will exist on its output $(V_J = +V_Y)$

(ii) In Fig. 9.4(b), the switch $S_2$ is closed, the op-amp $OA_6$ along with resistors $R_8$ will work as an inverting amplifier and $+V_Y$ will exist on its output $(V_J = +V_Y)$.

During the OFF time $T_1$ of this rectangular wave $V_M$

(iii) In Fig. 9.4(a), the switch $S_2$ is opened, the op-amp $OA_6$ along with resistors $R_8$ will work as inverting amplifiers and $-V_Y$ will exist on its output $(V_J = -V_Y)$.

(iv) In Fig. 9.4(b), the switch $S_2$ is opened, the op-amp $OA_6$ along with resistors $R_8$ will work as non-inverting amplifiers and $-V_Y$ will exist on its output $(V_J = -V_Y)$.

Another rectangular asymmetrical wave $V_J$ with peak-to-peak values of $\pm V_Y$ is generated at the output of op-amp $OA_6$. The $R_9C_2$ low pass filter gives the average value of the pulse train $V_J$ and is given as

$$V_O = \frac{1}{T}\left[\int_0^{T_4} V_Y\, dt + \int_{T_4}^{T_3+T_4}(-V_Y)\,dt\right] = \frac{V_Y}{T}(T_4 - T_3) \quad (9.12)$$

Triangular Wave-Referenced MCD

$$V_O = \frac{V_Y V_3}{V_T} \tag{9.13}$$

Equation (9.10) in (9.13) gives

$$V_O = \frac{V_2 V_3}{V_1} \tag{9.14}$$

Design exercise:

1. In the MCD circuits shown in Figs. 9.4(a) and (b), the triangular wave generator shown in Fig. 6.1(a) is used. Replace this triangular wave generator with the other triangular wave generator shown in Fig. 6.1(b). (i) Draw circuit diagrams, (ii) draw waveforms at appropriate places, (iii) explain the working operation, and (iv) deduce the expression for their output voltages.
2. The switches $S_1$ and $S_2$ in Figs. 9.4(a),(b) can be replaced with transistor switches as shown in Figs. 9.6(a) and (b). In each, (i) draw waveforms at the appropriate places, (ii) explain working operation, and (iii) deduce the expression for their output voltages.
3. Replace the switches $S_1$ and $S_2$ in Figs. 9.4(a) and (b) with FET switches and MOSFET switches. In each, (i) draw circuit diagrams, (ii) draw waveforms at appropriate places, (iii) explain the working operation, and (iv) deduce expression for their output voltages.

## 9.3 MULTIPLY–DIVIDE TIME-DIVISION MCD

The circuit diagrams of triangular wave-based multiply–divide time-division MCDs are shown in Fig 9.7 and their associated waveforms are shown in Fig. 9.8. Fig. 9.7(a) shows series-switching MCD and Fig. 9.7(b) shows shunt-switching MCD. A triangular wave $V_{T1}$ with a $\pm V_T$ peak-to-peak value is generated by the op-amp $OA_1$ transistors $Q_1$ and $Q_2$.

The second input voltage $V_2$ is compared with the generated triangular wave $V_{T1}$ by the comparator $OA_2$. An asymmetrical rectangular waveform $V_M$ is generated at the comparator $OA_2$ output. From the waveforms shown in Fig. 9.8, it is observed that

$$T_1 = \frac{V_T - V_2}{2V_T}T, \quad T_2 = \frac{V_T + V_2}{2V_T}T, \quad T = T_1 + T_2 \tag{9.15}$$

This rectangular wave $V_M$ is given as control input to the switch $S_1$.

**124**  Multiplier-Cum-Divider Circuits

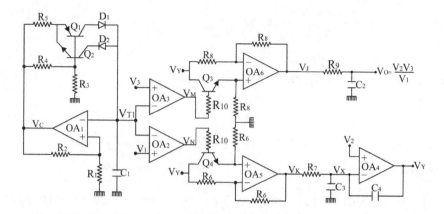

**FIGURE 9.6(A)**  Series-switching divide–multiply time-division MCD

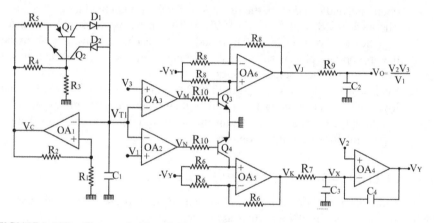

**FIGURE 9.6(B)**  Shunt-switching divide–multiply time-division MCD

During ON time $T_2$ of this rectangular wave $V_M$

(i) In Fig. 9.7(a), the switch $S_1$ is closed, the op-amp $OA_4$ along with resistors $R_6$ will work as a non-inverting amplifier and $+V_3$ will exist on its output ($V_J = +V_3$).

(ii) In Fig. 9.7(b), the switch $S_1$ is closed, the op-amp $OA_4$ along with resistors $R_6$ will work as an inverting amplifier and $+V_3$ will exist on its output ($V_J = +V_3$).

During OFF time $T_1$ of this rectangular wave $V_M$

(iii) In Fig. 9.7(a), the switch $S_1$ is opened, the op-amp $OA_4$ along with resistors $R_6$ will work as an inverting amplifier and $-V_3$ will exist on its output ($V_J = -V_3$).

# Triangular Wave-Referenced MCD

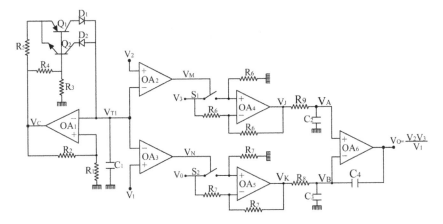

**FIGURE 9.7(A)** Series-switching multiply–divide time-division MCD

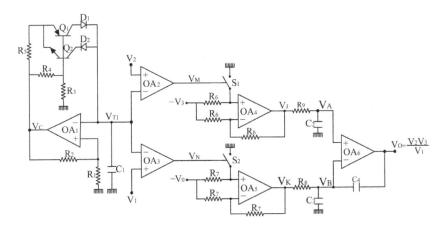

**FIGURE 9.7(B)** Shunt-switching multiply–divide time-division MCD

(iv) In Fig. 9.7(b), the switch $S_1$ is opened, the op-amp $OA_4$ along with resistors $R_6$ will work as a non-inverting amplifier and $-V_3$ will exist on its output ($V_J = -V_3$).

Another rectangular asymmetrical wave $V_J$ with peak-to-peak values is $\pm V_3$, which is generated at the output of op-amp $OA_4$. The $R_9 C_2$ low pass filter gives the average value of the pulse train $V_J$ and is given as

$$V_A = \frac{1}{T}\left[\int_O^{T_2} V_3\, dt + \int_{T_2}^{T_1+T_2} (-V_3)\, dt\right] = \frac{V_3}{T}(T_2 - T_1)$$

$$V_A = \frac{V_2 V_3}{V_T} \tag{9.16}$$

# Multiplier-Cum-Divider Circuits

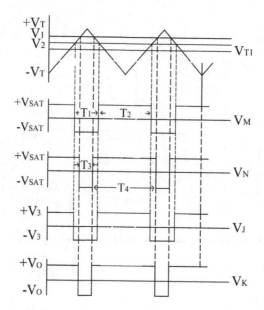

**FIGURE 9.8** Associated waveforms of Fig. 9.7

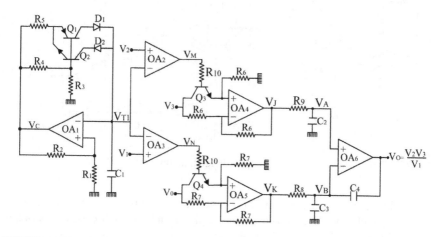

**FIGURE 9.9(A)** Series switching multiply–divide time-division MCD

The first input voltage $V_1$ is compared with the generated triangular wave $V_{T1}$ by the comparator on $OA_3$. An asymmetrical rectangular waveform $V_N$ is generated at the comparator $OA_3$ output. From the waveforms shown in Fig. 9.8, it is observed that

$$T_3 = \frac{V_T - V_1}{2V_T}T, \quad T_4 = \frac{V_T + V_1}{2V_T}T, \quad T = T_3 + T_4 \qquad (9.17)$$

# Triangular Wave-Referenced MCD

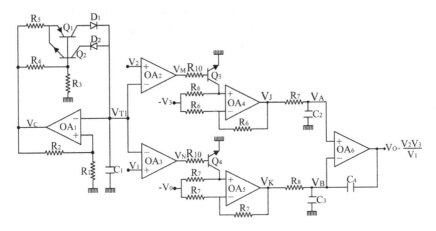

**FIGURE 9.9(B)** Shunt switching multiply–divide time-division MCD

This rectangular wave $V_N$ is given as control input to the switch $S_2$.
During the ON time $T_4$ of this rectangular wave $V_N$

(i) In Fig. 9.7(a), the switch $S_2$ is closed, the op-amp $OA_5$ along with resistors $R_7$ will work as a non-inverting amplifier and $+V_O$ will exist on its output ($V_K = +V_O$).
(ii) In Fig. 9.7(b), the switch $S_2$ is closed, the op-amp $OA_5$ along with resistors $R_7$ will work as an inverting amplifier and $+V_O$ will exist on its output ($V_K = +V_O$).

During the OFF time $T_3$ of this rectangular wave $V_N$

(iii) In Fig. 9.7(a), the switch $S_2$ is opened, the op-amp $OA_5$ along with resistors $R_7$ will work as an inverting amplifier and $-V_O$ will exist on its output ($V_K = -V_O$).
(iv) In Fig. 9.7(b), the switch $S_2$ is opened, the op-amp $OA_5$ along with resistors $R_7$ will work as non-inverting amplifier and $-V_O$ will exist on its output ($V_K = -V_O$).

Another rectangular asymmetrical wave $V_K$ with peak-to-peak values of $\pm V_O$ is generated at the output of op-amp $OA_5$. The $R_8C_3$ low pass filter gives the average value of the pulse train $V_K$ and is given as

$$V_B = \frac{1}{T}\left[\int_0^{T_4} V_O\, dt + \int_{T_4}^{T_3+T_4} (-V_O)\, dt\right] = \frac{V_O}{T}(T_4 - T_3)$$

$$V_B = \frac{V_1 V_O}{V_T} \tag{9.18}$$

The op-amp $OA_6$ is configured in a negative closed loop feedback and a positive DC voltage is ensured in the feedback loop. Hence its inverting terminal voltage must be equal to its non-inverting terminal voltage, i.e.

$$V_A = V_B \tag{9.19}$$

From equations (9.16) and (9.18)

$$V_O = \frac{V_2 V_3}{V_1} \tag{9.20}$$

Design exercise:

1. In the MCD circuits shown in Figs. 9.7(a) and (b), the triangular wave generator shown in Fig. 6.1(a) is used. Replace this triangular wave generator with the other triangular wave generator shown in Fig. 6.1(b). (i) Draw circuit diagrams, (ii) draw waveforms at appropriate places, (iii) explain the working operation, and (iv) deduce the expression for their output voltages.
2. The switches $S_1$ and $S_2$ in Figs. 9.7(a) and (b) can be replaced with transistor switches and shown in Figs. 9.9(a) and (b). (i) Draw waveforms at appropriate places, (ii) explain the working operation, and (iii) deduce the expression for their output voltages.
3. Replace the switches $S_1$ and $S_2$ in Figs. 9.7(a) and (b) with FET switches and MOSFET switches. In each, (i) draw circuit diagrams, (ii) draw waveforms at appropriate places, (iii) explain the working operation, and (iv) deduce the expression for their output voltages.

# 10 Peak-Responding MCDs with Analog Switches

Peak-responding MCDs are classified as (i) peak-detecting MCDs and (ii) peak-sampling MCDs. A short pulse / sawtooth waveform whose time period T is (i) proportional to one voltage ($-V_2$) and (ii) inversely proportional to one voltage ($V_1$) is generated. The third input voltage $V_3$ is integrated during this time period T. The peak value of the integrated output is proportional to $\dfrac{V_2 V_3}{V_1}$. This is called a double single-slope peak-responding MCD. A square wave / triangular waveform whose time period T is (i) proportional to one voltage ($-V_2$) and (ii) inversely proportional to one voltage ($V_1$) is generated. The third input voltage $V_3$ is integrated during this time period T. The peak value of the integrated output is proportional to $\dfrac{V_2 V_3}{V_1}$. This is called a double dual-slope peak-responding MCD. A rectangular waveform whose (i) time period T is inversely proportional to first input voltage $V_1$ and (ii) OFF time proportional to second input voltage ($V_2$) is generated. The third input voltage $V_3$ is integrated during this OFF time. The peak value of the integrated output is proportional to $\dfrac{V_2 V_3}{V_1}$. This is called a pulse-width integrated peak-responding MCD.

At the output of a peak-responding MCD, if a peak detector is used, it is called a peak-detecting MCD and if sample and hold is used, it is called a peak-sampling MCD. The peak-responding MCDs of both peak-detecting MCDs and peak-sampling MCDs using analog switches are described in this chapter.

## 10.1 DOUBLE SINGLE-SLOPE PEAK-RESPONDING MCDS

The circuit diagrams of double single-slope peak-responding MCDs are shown in Fig. 10.1 and their associated waveforms are shown in Fig. 10.2. Fig. 10.1(a) shows a double single-slope peak-detecting multiplier and Fig. 10.1(b) shows a double single-slope peak-sampling multiplier. Let's initially assume the comparator $OA_2$ output is

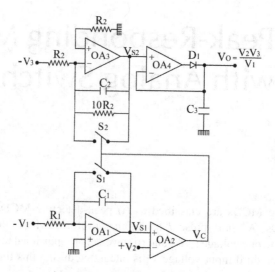

**FIGURE 10.1(A)** Switching double single-slope peak-detecting multiplier

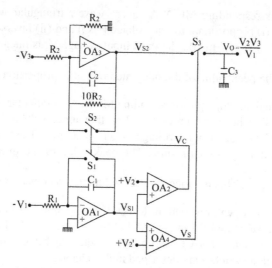

**FIGURE 10.1(B)** Switching double single-slope peak-sampling multiplier

LOW, the switch $S_1$ is open, and an integrator formed by resistor $R_1$, capacitor $C_1$, and op-amp $OA_1$ integrates the input voltage $(-V_1)$. The integrated output will be

$$V_{S1} = -\frac{1}{R_1 C_1}\int -V_1 dt = \frac{V_1}{R_1 C_1} t \qquad (10.1)$$

A positive-going ramp $V_{S_1}$ is generated at the output of op-amp $OA_1$. When the output of $OA_1$ reaches the voltage level of $V_2$, the comparator $OA_2$ output becomes HIGH.

# Peak-Responding MCD with Analog Switches

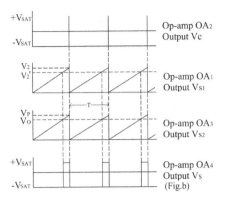

**FIGURE 10.2** Associated waveforms of Fig. 10.1

The switch $S_1$ is closed and hence the capacitor $C_1$ is shorted so that op-amp $OA_1$ output becomes ZERO. Then op-amp $OA_2$ output goes to LOW, the switch $S_1$ opens and the integrator composed by $R_1$, $C_1$ and op-amp $OA_1$ integrates the input voltage $(-V_1)$ and the cycle therefore repeats to provide (i) a sawtooth wave of peak value $V_2$ at the output of op-amp $OA_1$ and (ii) a short pulse waveform $V_C$ at the output of comparator $OA_2$. The short pulse $V_C$ also controls switch $S_2$. During the short HIGH time of $V_C$, switch $S_2$ is closed and the capacitor $C_2$ is short circuited so that the op-amp $OA_3$ output is zero volts. During the LOW time of $V_C$, switch $S_2$ opens, the integrator formed by resistor $R_2$, capacitor $C_2$, and op-amp $OA_3$ integrates its input voltage $(-V_3)$, and its output is given as

$$V_{S2} = -\frac{1}{R_2 C_2}\int -V_3 dt = \frac{V_3}{R_2 C_2} t \qquad (10.2)$$

Another sawtooth waveform $V_{S2}$ with the peak value $V_P$ is generated at the output of op-amp $OA_3$. From the waveforms shown in Fig. 10.2 and from equations (10.1), (10.2) and the fact that at $t = T$, $V_{S1} = V_2$, $V_{S2} = V_P$ we get.

$$V_2 = \frac{V_1}{R_1 C_1} T \qquad (10.3)$$

$$V_P = \frac{V_3}{R_2 C_2} T \qquad (10.4)$$

From equations (10.3) and (10.4)

$$V_P = \frac{V_3}{R_2 C_2}\frac{V_2}{V_1} R_1 C_1$$

## Multiplier-Cum-Divider Circuits

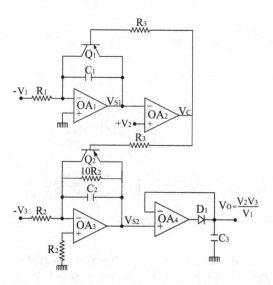

**FIGURE 10.3(A)** Switching double single-slope peak-detecting multiplier

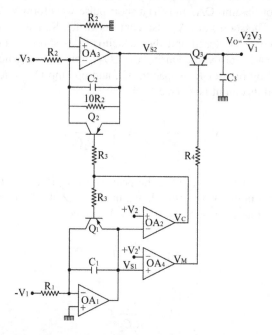

**FIGURE 10.3(B)** Switching double single-slope peak-sampling multiplier

Peak-Responding MCD with Analog Switches

Let us assume $R_1 = R_2$ and $C_1 = C_2$, then

$$V_P = \frac{V_1 V_2}{V_R} \qquad (10.5)$$

(i) In the circuit shown in Fig. 10.1(a), the peak detector realized by op-amp $OA_4$, diode $D_1$, and capacitor $C_3$ gives this peak value $V_P$ at its output $V_O$. $V_O = V_P$.

(ii) In the circuit shown in Fig. 10.1(b), the peak value VP is obtained by the sample and hold circuit realized by switch $S_3$ and capacitor $C_3$. The sampling pulse is generated by op-amp $OA_4$ by comparing a voltage $V_2'$, which is slightly less than $V_2$, called with the sawtooth wave $VS_1$. The sample and hold operation is illustrated graphically in Fig. 10.2. The sample and hold output $V_O = V_P$.

From equation (10.5), the output will be $V_O = V_P$

$$V_O = \frac{V_1 V_2}{V_R} \qquad (10.6)$$

Design exercise:

1. The switches in Figs. 10.1(a) and (b) can be replaced with transistorized switches as shown in Figs. 10.3(a) and (b) respectively. (i) Explain the working operation of the multipliers shown in Figs. 10.3(a) and (b), (ii) draw the waveforms at appropriate places, and (iii) deduce the expression for their output voltages.
2. The switches in Figs. 10.1(a) and (b) are to be replaced with FET switches and MOSFET switches. In each, (i) draw the circuit diagrams, (ii) explain their working operations, (iii) draw waveforms at appropriate places and (iv) Deduce expression for their output voltages

## 10.2 DOUBLE DUAL-SLOPE MCD USING FEEDBACK COMPARATOR

The MCDs using the double dual-slope principle are shown in Fig 10.4 and their associated waveforms are shown in Fig 10.5. Fig. 10.4(a) shows a series-switching peak-detecting MCD, Fig. 10.4(b) shows a shunt-switching peak-detecting MCD, Fig. 10.4(c) shows a series-switching peak-sampling MCD and Fig. 10.4(d) shows a shunt-switching peak-sampling MCD. Let initially the comparator $OA_2$ output be LOW.

(i) In Figs 10.4(a) and (c), the switch $S_1$ is opened, op-amp $OA_6$ along with resistors $R_5$ will work as an inverting amplifier and $(-V_1)$ will be at its output $(V_M = -V_1)$. The switch $S_2$ is opened, op-amp $OA_7$ along with resistors $R_6$

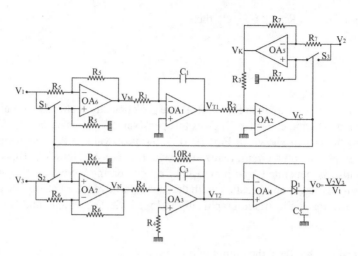

**FIGURE 10.4(A)** Series-switching double dual-slope peak-detecting MCD

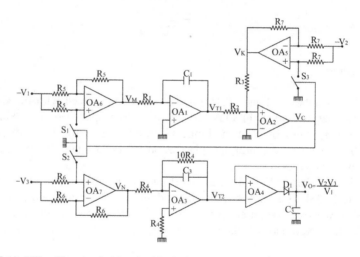

**FIGURE 10.4(B)** Shunt-switching double dual-slope peak-detecting MCD

will work as an inverting amplifier and $(-V_3)$ will be its output $(V_N = -V_3)$. The switch $S_3$ is opened, op-amp $OA_5$ along with resistors $R_7$ will work as inverting amplifier and $(-V_2)$ will be at its output $(V_K = -V_2)$.

(ii) In Figs 10.4(b) and (d), the switch $S_1$ is opened, op-amp $OA_6$ along with resistors $R_5$ will work as a non-inverting amplifier and $(-V_1)$ will be at its output $(V_M = -V_1)$. The switch $S_2$ is opened, op-amp $OA_7$ along with resistors $R_6$ will work as a non-inverting amplifier and $(-V_3)$ will be its output $(V_N = -V_3)$. The switch $S_3$ is opened, op-amp $OA_5$ along with resistors $R_7$ will work as a non-inverting amplifier and $(-V_2)$ will be at its output $(V_K = -V_2)$.

## Peak-Responding MCD with Analog Switches

The output of integrator composed by resistor $R_1$, capacitor $C_1$, and op-amp $OA_1$ will be

$$V_{T1} = -\frac{1}{R_1 C_1}\int(-V_1)dt = \frac{V_1}{R_1 C_1}t \qquad (10.7)$$

The output of integrator $OA_1$ is going towards positive saturation and when it reaches a value '$+V_T$', the comparator $OA_2$ output becomes HIGH.

(i) In Figs 10.4(a) and (c), the switch $S_1$ is closed, op-amp $OA_6$ along with resistors $R_5$ will work as a non-inverting amplifier and $(+V_1)$ will be at its output $(V_M = +V_1)$. The switch $S_2$ is closed, op-amp $OA_7$ along with resistors $R_6$ will work as a non-inverting amplifier and $(+V_3)$ will be its output $(V_N = +V_3)$. The switch $S_3$ is closed, op-amp $OA_5$ along with resistors $R_7$ will work as non-inverting amplifier and $(+V_2)$ will be at its output $(V_K = +V_2)$

(ii) In Figs 10.4(b) and (d), the switch $S_1$ is closed, op-amp $OA_6$ along with resistors $R_5$ will work as an inverting amplifier and $(+V_1)$ will be at its output $(V_M = +V_1)$. The switch $S_2$ is closed, op-amp $OA_7$ along with resistors $R_6$ will work as an inverting amplifier, and $(+V_3)$ will be its output $(V_N = +V_3)$. The switch $S_3$ is closed, op-amp $OA_5$ along with resistors $R_7$ will work as inverting amplifier, and $(+V_2)$ will be at its output $(V_K = +V_2)$.

The integrator $OA_1$ output will now be

$$V_{T1} = -\frac{1}{R_1 C_1}\int(+V_1)dt = -\frac{V_1}{R_1 C_1}t \qquad (10.8)$$

The output of integrator $OA_1$ is reversing towards negative saturation and when it reaches a value '$(-V_T)$,' the comparator $OA_2$ output becomes LOW. The switch $S_1$ selects $(-V_1)$ and the sequence repeats to give (i) a triangular waveform $V_{T1}$ of $\pm V_T$ peak-to-peak values with a time period of 'T,' (ii) square waveform $V_C$ at the output of comparator $OA_2$, (iii) second square waveform $V_M$ with $\pm V_1$ peak-to-peak value at the output of op-amp $OA_6$, (iv) third square waveform $V_N$ with $\pm V_3$ peak-to-peak values at the output of op-amp $OA_7$, and (v) fourth square waveform $V_K$ with $\pm V_2$ peak-to-peak values at the output of op-amp $OA_5$. From the waveforms shown in Fig.10.5, equation (10.7) and the fact that at $t = T/2$, $V_{T1} = 2V_T$ we get

$$2V_T = \frac{V_1}{R_1 C_1}\frac{T}{2}$$

$$T = \frac{4V_T}{V_1}R_1 C_1 \qquad (10.9)$$

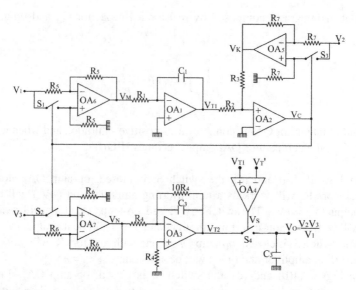

**FIGURE 10.4(C)** Series-switching double dual-slope peak-sampling MCD

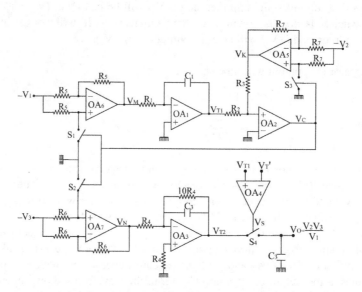

**FIGURE 10.4(D)** Shunt-switching double dual-slope peak-sampling MCD

When the comparator $OA_2$ output is LOW ($-V_{SAT}$), the effective voltage at the non-inverting terminal of comparator $OA_2$ will be by superposition principle

$$\frac{(-V_2)}{(R_2+R_3)}R_2 + \frac{(+V_T)}{(R_2+R_3)}R_3$$

# Peak-Responding MCD with Analog Switches

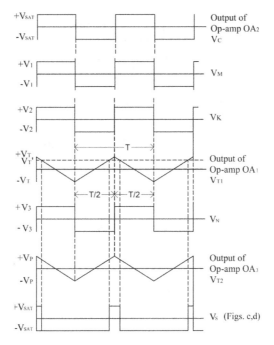

**FIGURE 10.5** Associated waveforms of Fig. 10.4

When this effective voltage becomes zero

$$\frac{(-V_2)R_2 + (+V_T)R_3}{(R_2 + R_3)} = 0$$

$$(+V_T) = (+V_2)\frac{R_2}{R_3}$$

When the comparator $OA_2$ output is HIGH ($+V_{CC}$), the effective voltage at the non-inverting terminal of comparator $OA_2$ will be by superposition principle

$$\frac{(+V_2)}{(R_2 + R_3)}R_2 + \frac{(-V_T)}{(R_2 + R_3)}R_3$$

When this effective voltage becomes zero

$$\frac{(+V_2)R_2 + (-V_T)R_3}{(R_2 + R_3)} = 0$$

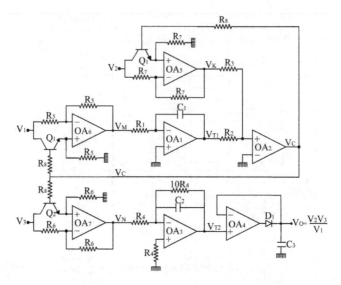

**FIGURE 10.6(A)** Series-switching double dual-slope peak-detecting MCD

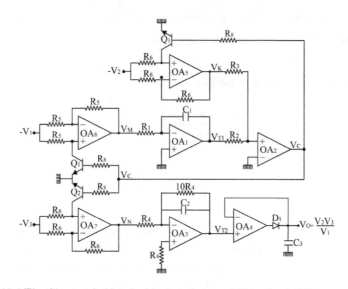

**FIGURE 10.6(B)** Shunt-switching double dual-slope peak-detecting MCD

$$(-V_T) = (-V_2)\frac{R_2}{R_3}$$

$$\pm V_T = \pm \frac{R_2}{R_3}V_2 \tag{10.10}$$

## Peak-Responding MCD with Analog Switches

The third square wave $V_N$ is converted into a triangular wave $V_{T2}$ by the integrator II composed by resistor $R_4$, capacitor $C_2$, and op-amp $OA_3$ with $\pm V_P$ as peak-to-peak values of the same time period T. For one transition the integrator II output is given as

$$V_{T2} = -\frac{1}{R_4 C_2} \int (-V_3) dt = \frac{V_3}{R_4 C_2} t \qquad (10.11)$$

From the waveforms shown in Fig. 10.5, the equation (10.11), and the fact that at $t = T/2$, $V_{T2} = 2V_p$ we get

$$2V_P = \frac{V_3}{R_4 C_2} \frac{T}{2}$$

$$V_P = \frac{V_2 V_3}{V_1} \frac{R_2}{R_3} \frac{R_1 C_1}{R_4 C_2}$$

Let $R_1 = R_4$ and $C_1 = C_2$ and $R_2 = R_3$ (but in practical $R_2 < R_3$)

$$V_P = \frac{V_2 V_3}{V_1}$$

(i) In the MCD circuit shown in Figs. 10.4(a) and (b), the peak detector realized by op-amp $OA_4$, diode $D_1$, and capacitor $C_3$ gives the peak value '$V_P$' of the triangular wave $V_{T2}$ and hence $V_O = V_P$.

(ii) In the MCD circuits shown in Figs. 10.4(c) and (d), the peak value $V_P$ of the triangular waveform $V_{T2}$ is obtained by the sample and hold circuit realized by switch $S_4$ and capacitor $C_3$. The sampling pulse VS is generated by op-amp $OA_4$ by comparing a voltage $V_T'$, which is slightly less than $V_T$, with the triangular wave of $V_{T1}$. The sample and hold operation is illustrated graphically in Fig. 10.5. The sampled output is given as $V_O = V_P$.

$$V_O = \frac{V_2 V_3}{V_1} \qquad (10.12)$$

Design exercise:

1. The switches in Figs. 10.4(a) and (b) can be replaced with transistorized switches and shown in Figs. 10.6(a) and (b) respectively. (i) Explain the working operation of the multipliers shown in Figs. 10.6(a) and (b), (ii) draw the waveforms at appropriate places, and (iii) deduce the expression for their output voltages.

2. The switches in Figs. 10.4(a) and (d) are to be replaced with FET switches and MOSFET switches. In each, (i) draw the circuit diagrams, (ii) explain their working operations, (iii) draw waveforms at appropriate places, and (iv) deduce the expression for their output voltages.

## 10.3 DOUBLE DUAL-SLOPE MCD WITH FLIP FLOP

The MCDs using the double dual-slope principle with flip flop are shown in Fig 10.7 and their associated waveforms are shown in Fig 10.8. Fig. 10.7(a) shows a series switching peak-detecting MCD, Fig. 10.7(b) shows a shunt-switching peak-detecting MCD, Fig. 10.7(c) shows a series-switching peak-sampling MCD and Fig. 10.7(d) shows a shunt-switching peak-sampling MCD. Let's initially assume the flip flop output $V_C$ is LOW.

(i) In Figs 10.7(a) and (c), the switch $S_1$ is opened, op-amp $OA_6$ along with resistors $R_5$ will work as an inverting amplifier and $(-V_1)$ will be at its output

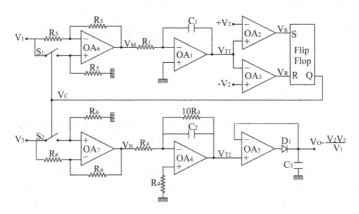

**FIGURE 10.7(A)**   Series-switching double dual-slope peak-detecting MCD with flip flop

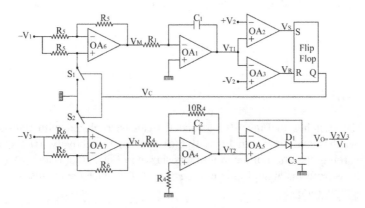

**FIGURE 10.7(B)**   Shunt-switching double dual-slope peak-detecting MCD with flip flop

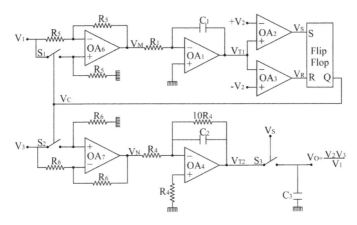

**FIGURE 10.7(C)** Series-switching double dual-slope peak-sampling MCD with flip flop

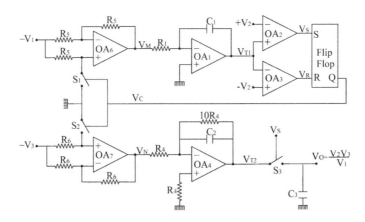

**FIGURE 10.7(D)** Shunt-switching double dual-slope sampling MCD with flip flop

($V_M = -V_1$). The switch $S_2$ is opened, op-amp $OA_7$ along with resistors $R_6$ will work as an inverting amplifier and ($-V_3$) will be at its output ($V_N = -V_3$).

(ii) In Figs 10.7(b) and (d), the switch $S_1$ is opened, op-amp $OA_6$ along with resistors $R_5$ will work as a non-inverting amplifier and ($-V_1$) will be at its output ($V_M = -V_1$). The switch $S_2$ is opened, op-amp $OA_7$ along with resistors $R_6$ will work as non-inverting amplifier and ($-V_3$) will be at its output ($V_N = -V_3$).

The output of integrator I composed by resistor $R_1$, capacitor $C_1$, and op-amp $OA_1$ will be

$$V_{T1} = -\frac{1}{R_1 C_1}\int(-V_1)dt = \frac{V_1}{R_1 C_1}t \qquad (10.13)$$

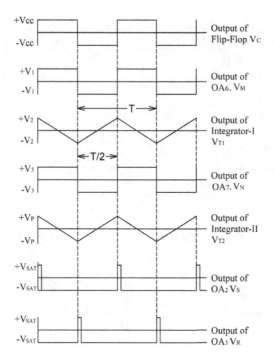

**FIGURE 10.8** Associated waveforms of Fig. 10.7

The output of integrator I is going towards positive saturation and when it reaches the value '+$V_2$', the comparator $OA_2$ output becomes HIGH and it sets the flip flop output to HIGH.

(i) In Figs 10.7(a) and (c), the switch $S_1$ is closed, op-amp $OA_6$ along with resistors $R_5$ will work as a non-inverting amplifier, and (+$V_1$) will be at its output ($V_M = +V_1$). The switch $S_2$ is closed, op-amp $OA_7$ along with resistors $R_6$ will work as a non-inverting amplifier and (+$V_3$) will be at its output ($V_N = +V_3$).

(ii) In Figs 10.7(b) and (d), the switch $S_1$ is closed, op-amp $OA_6$ along with resistors $R_5$ will work as an inverting amplifier and (+$V_1$) will be at its output ($V_M = +V_1$). The switch $S_2$ is closed, op-amp $OA_7$ along with resistors $R_6$ will work as an inverting amplifier and (+$V_3$) will be at its output ($V_N = +V_3$).

The output of integrator I composed by resistor $R_1$, capacitor $C_1$, and op-amp $OA_1$ will now be

$$V_{T1} = -\frac{1}{R_1 C_1}\int (+V_1)dt = -\frac{V_1}{R_1 C_1}t \qquad (10.14)$$

The output of integrator $OA_1$ is reversing towards negative saturation and when it reaches the value '(−$V_2$),' the comparator $OA_3$ output becomes HIGH and resets the

# Peak-Responding MCD with Analog Switches

flip flop so that its output $V_C$ becomes LOW and the sequence repeats to give (i) a triangular waveform $V_{T1}$ of $\pm V_2$ peak-to-peak value with a time period of 'T' at the output of integrator I, (ii) a square waveform $V_C$ at the output of flip flop, (iii) another square waveform $V_M$ with $\pm V_1$ peak-to-peak value at the output of op-amp $OA_6$, and (iv) third square waveform $V_N$ with $\pm V_3$ peak-to-peak value is generated at the output of op-amp $OA_7$. From the waveforms shown in Fig 10.8, equation 10.13 and the fact that at $t = T/2$, $V_{T1} = 2V_2$ we get

$$2V_2 = \frac{V_1}{R_1 C_1} \frac{T}{2}$$

$$T = \frac{4V_2}{V_1} R_1 C_1$$

The third square wave $V_N$ is converted into triangular wave $V_{T2}$ by the integrator II composed by resistor $R_4$, capacitor $C_2$, and op-amp $OA_4$ with $\pm V_P$ as peak-to-peak values of same time period T. For one transition the integrator II output is given as

$$V_{T2} = -\frac{1}{R_4 C_2} \int (-V_3) dt = \frac{V_3}{R_4 C_2} t \qquad (10.15)$$

From the waveforms shown in Fig. 10.8 the equation (10.15), and the fact that at $t = T/2$, $V_{T2} = 2V_p$ we get

$$2V_P = \frac{V_3}{R_4 C_2} \frac{T}{2}$$

$$V_P = \frac{V_2 V_3}{V_1} \frac{R_1 C_1}{R_4 C_2}$$

Let $R_1 = R_4$ and $C_1 = C_2$

$$V_P = \frac{V_2 V_3}{V_1}$$

(i) In the MCD circuits shown in Figs 10.7(a) and (b), the peak detector realized by op-amp $OA_5$, diode $D_1$, and capacitor $C_3$, gives a peak value '$V_P$' of triangular wave $V_{T2}$ and hence $V_0 = V_P$.
(ii) In the MCD circuits shown in Figs. 10.7(c) and (d), the peak value $V_P$ of the triangular waveform $V_{T2}$ is obtained by the sample and hold circuit realized

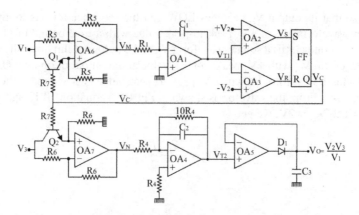

**FIGURE 10.9(A)** Series-switching double dual-slope peak-detecting MCD

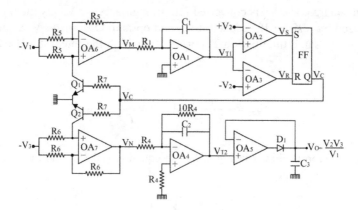

**FIGURE 10.9(B)** Shunt-switching double dual-slope peak-detecting MCD

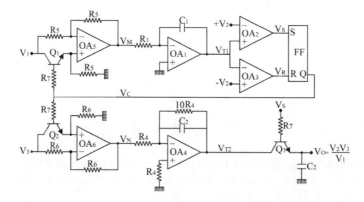

**FIGURE 10.9(C)** Series-switching double dual-slope peak-sampling MCD

## Peak-Responding MCD with Analog Switches

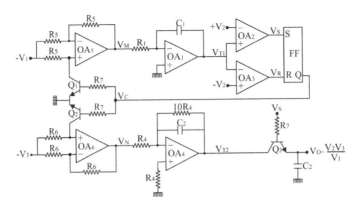

**FIGURE 10.9(D)** Shunt-switching double dual-slope peak-sampling MCD

by switch $S_3$ and capacitor $C_3$. The comparator $OA_2$ output VS is acting as a sampling pulse. The sample and hold operation is illustrated graphically in Fig. 10.8. The sampled output is given as $V_O = V_P$.

$$V_O = \frac{V_2 V_3}{V_1} \tag{10.16}$$

Design exercise:

1. The switches in Figs. 10.7(a) to (d) can be replaced with transistorized switches and shown in Figs. 10.9(a) to (d) respectively. (i) Explain the working operation of the multipliers shown in Figs. 10.9(a) to (d), (ii) draw waveforms at appropriate places, and (iii) deduce the expression for their output voltages.
2. The switches in Figs. 10.7(a) to (d) are to be replaced with FET switches and MOSFET switches. In each, (i) draw the circuit diagrams, (ii) explain their working operations, (iii) draw waveforms at appropriate places, and (iv) deduce the expression for their output voltages

## 10.4 PULSE-WIDTH INTEGRATED PEAK-RESPONDING MCD

The circuit diagrams of pulse-width integrated peak-responding MCDs are shown in Fig. 10.10 and their associated waveforms are shown in Fig. 10.11. Fig. 10.10(a) shows a pulse-width integrated peak-detecting MCD and Fig. 10.10(b) shows a pulse-width integrated peak-sampling MCD. Let's initially assume the comparator $OA_2$ output is LOW, the switch $S_1$ is open, and the integrator formed by resistor $R_1$, capacitor $C_1$, and op-amp $OA_1$ integrates $(-V_1)$. The integrated output is given as

$$V_{S1} = -\frac{1}{R_1 C_1} \int -V_1 dt = \frac{V_1}{R_1 C_1} t \tag{10.17}$$

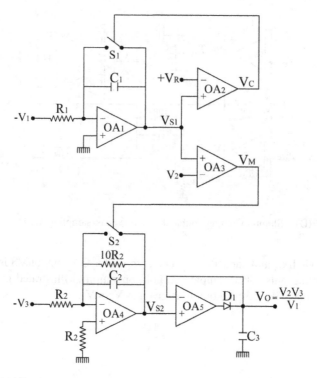

**FIGURE 10.10(A)** Pulse-width integrated peak-detecting MCD

When the output of op-amp $OA_1$ is rising towards positive saturation and it reaches the value '$V_R$,' the comparator $OA_2$ output will becomes HIGH, the switch $S_1$ is closed and the capacitor $C_1$ is short circuited, and the op-amp $OA_1$ output becomes zero. Now the comparator $OA_2$ output changes to LOW and the cycle therefore repeats to give (i) a sawtooth waveform $V_{S1}$ of peak value $V_R$ and time period T at the output of op-amp $OA_1$ and (ii) a short pulse waveform $V_C$ at the output of op-amp $OA_2$. From the waveforms shown in Fig.10.11, the fact that at t = T, $V_{S1} = V_R$ we get

$$V_R = \frac{V_1}{R_1 C_1} T, T = \frac{V_R}{V_1} R_1 C_1 \tag{10.18}$$

The sawtooth waveform $V_{S1}$ is compared with the second input voltage $V_2$ by the comparator $OA_3$. An asymmetrical rectangular wave $V_M$ is generated at the output of comparator $OA_3$. The OFF time of this wave is given as

$$\delta_T = \frac{V_2}{V_R} T \tag{10.19}$$

# Peak-Responding MCD with Analog Switches

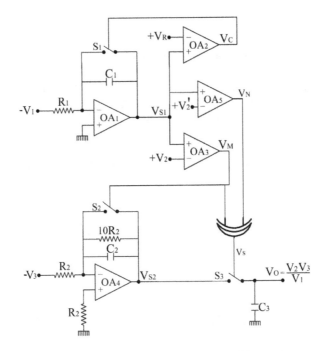

**FIGURE 10.10(B)** Pulse-width integrated peak-sampling MCD

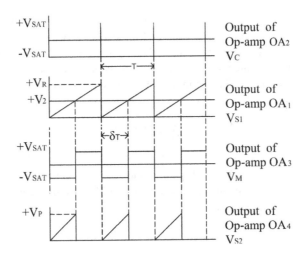

**FIGURE 10.11(A)** Associated waveforms of Fig. 10.10(a)

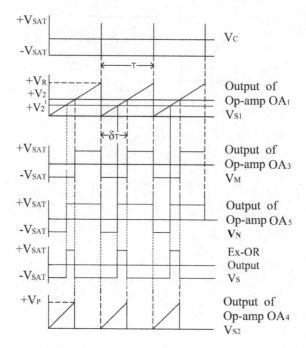

**FIGURE 10.11(B)**  Associated waveforms of Fig. 10.10(b)

The output of comparator OA$_3$ is given as the control input of switch S$_2$. During the ON time of V$_M$, the switch S$_2$ is closed and the capacitor C$_2$ is shorted so that zero volts appeared at the op-amp OA$_4$ output. During the OFF time of V$_M$, the switch S$_2$ is opened and another integrator is formed by resistor R$_2$, capacitor C$_2$, and op-amp OA$_4$. This integrator integrates the input voltage ($-V_3$) and its output is given as

$$V_{S2} = -\frac{1}{R_2 C_2}\int -V_3 dt = \frac{V_3}{R_2 C_2} t \tag{10.20}$$

A semi-sawtooth wave V$_{S2}$ with peak values of V$_P$ is generated at the output of op-amp OA$_4$. From the waveforms shown in Fig.10.11, from the equation (10.20) and fact that at t = $\delta_T$, V$_{S2}$ = V$_P$ web get

$$V_P = \frac{V_3}{R_2 C_2}\delta_T$$

$$V_P = \frac{V_2 V_3}{V_1}\frac{R_1 C_1}{R_2 C_2}$$

# Peak-Responding MCD with Analog Switches

Let us assume $R_1 = R_2$ and $C_1 = C_2$

$$V_P = \frac{V_2 V_3}{V_1} \qquad (10.21)$$

(i) In Fig. 10.10(a), the peak detector realized by op-amp $OA_5$, diode $D_1$, and capacitor $C_3$ gives this peak value $V_P$ at its output. $V_0 = V_P$

(ii) In the circuit shown in Figs. 10.10(b), the peak value $V_P$ is obtained by the sample and hold circuit realized by switch $S_3$ and capacitor $C_3$. The sampling pulse $V_S$ is generated by the Ex-OR gate from the signals $V_M$ and $V_N$. $V_N$ is obtained by comparing a voltage $V_2'$ which is slightly less than $V_2$, i.e. $V_2'$ with the sawtooth waveform $V_{S1}$. The sampled output is given as $V_0 = V_P$.

From equation (10.21), the output voltage will be

$$V_O = \frac{V_2 V_3}{V_1} \qquad (10.22)$$

Design exercise:

1. The switches in Figs. 10.10(a) and (b) can be replaced with transistorized switches as shown in Figs. 10.12(a) and (b). (i) Explain the working operation of the multipliers shown in Figs. 10.12(a) and (b), (ii) draw the waveforms at appropriate places, and (iii) deduce the expression for their output voltages.
2. The switches in Fig. 10.10 are to be replaced with FET switches and MOSFET switches. In each, (i) draw the circuit diagrams, (ii) explain their working operations, (iii) draw the waveforms at appropriate places, and (iv) deduce the expression for their output voltages.

## 10.5 MCD USING VOLTAGE-TUNABLE ASTABLE MULTIVIBRATOR

The MCDs using a voltage-tunable astable multivibrator are shown in Fig. 10.13 and their associated waveforms are shown in Fig. 10.14. Fig. 10.13(a) shows a series-switching peak-detecting MCD, Fig. 10.13(b) shows a shunt-switching peak-detecting MCD, Fig. 10.13(c) shows series-switching peak-sampling MCD, and Fig. 10.13(d) shows a shunt-switching peak-sampling MCD. The op-amp $OA_1$ in association with the components $R_1$, $R_2$, $R_3$, and $C_1$ produces a square waveform $V_C$. The time period of this square waveform $V_C$ is proportional to the input voltage $V_2$ and inversely proportional to another input voltage $V_1$.

$$T = K \frac{V_2}{V_1} \qquad (10.23)$$

# Multiplier-Cum-Divider Circuits

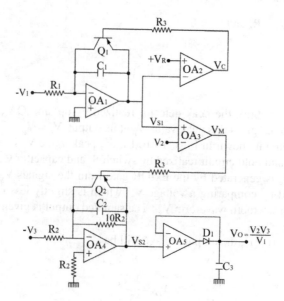

**FIGURE 10.12(A)** Pulse-width integrated peak-detecting MCD

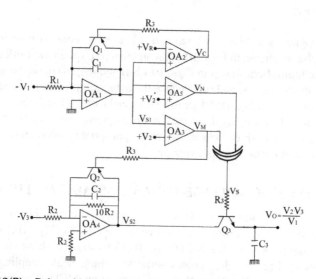

**FIGURE 10.12(B)** Pulse-width integrated peak-sampling MCD

# Peak-Responding MCD with Analog Switches

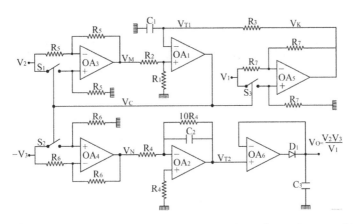

**FIGURE 10.13(A)** Series-switching peak-detecting MCD using voltage-tunable astable multivibrator

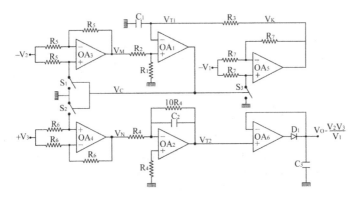

**FIGURE 10.13(B)** Shunt-switching peak-detecting MCD using voltage-tunable astable multivibrator

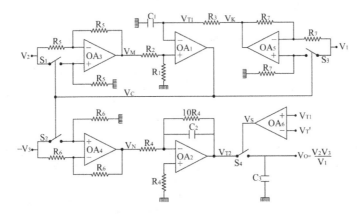

**FIGURE 10.13(C)** Series-switching peak-sampling MCD using voltage-tunable astable multivibrator

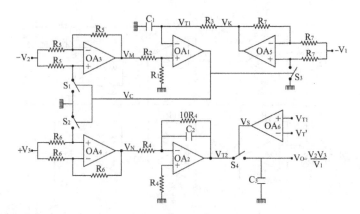

**FIGURE 10.13(D)** Shunt-switching peak-sampling MCD using voltage-tunable astable multivibrator

During the HIGH value of $V_C$,

(i) In Figs 10.13(a) and (c), the switch $S_1$ is closed, op-amp $OA_3$ along with resistors $R_5$ will work as a non-inverting amplifier and $(+V_2)$ will be at its output $(V_M = +V_2)$. The switch $S_2$ is closed, op-amp $OA_4$ along with resistors $R_6$ will work as a non-inverting amplifier and $(-V_3)$ will be at its output $(V_N = -V_3)$. The switch $S_3$ is closed, op-amp $OA_5$ along with resistors $R_7$ will work as non-inverting amplifier and $(+V_1)$ will be at its output $(V_K = +V_1)$.

(ii) In Figs. 10.12(b) and (d), the switch $S_1$ is closed, op-amp $OA_3$ along with resistors $R_5$ will work as an inverting amplifier and $(+V_2)$ will be at its output $(V_M = +V_2)$. The switch $S_2$ is closed, op-amp $OA_4$ along with resistors $R_6$ will work as an inverting amplifier and $(-V_3)$ will be at its output $(V_N = -V_3)$. The switch $S_3$ is closed, op-amp $OA_5$ along with resistors $R_7$ will work as an inverting amplifier and $(+V_1)$ will be at its output $(V_K = +V_1)$.

During the LOW value of $V_C$

(i) In Figs 10.13(a) and (c), the switch $S_1$ is open, op-amp $OA_3$ along with resistors $R_5$ will work as an inverting amplifier and $(-V_2)$ will be at its output $(V_M = -V_2)$. The switch $S_2$ is opened, op-amp $OA_4$ along with resistors $R_6$ will work as an inverting amplifier and $(+V_3)$ will be at its output $(V_N = +V_3)$. The switch $S_3$ is opened, op-amp $OA_5$ along with resistors $R_7$ will work as an inverting amplifier and $(-V_1)$ will be at its output $(V_K = -V_1)$.

(ii) In Figs. 10.13(b) and (d), the switch $S_1$ is open, op-amp $OA_3$ along with resistors $R_5$ will work as a non-inverting amplifier and $(-V_2)$ will be at its output $(V_M = -V_2)$. The switch $S_2$ is opened, op-amp $OA_4$ along with resistors $R_6$ will work as a non-inverting amplifier and $(+V_3)$ will be at its output $(V_N = +V_3)$. The switch $S_3$ is opened, op-amp $OA_5$ along with resistors $R_7$ will work as non-inverting amplifier and $(-V_1)$ will be at its output $(V_M = -V_1)$

## Peak-Responding MCD with Analog Switches

Three square waves are generated: (i) $V_M$ with a $\pm V_2$ peak-to-peak value at the output of op-amp $OA_3$, (ii) $V_N$ with a $\pm V_3$ peak-to-peak value at the output of op-amp $OA_4$, and (iii) $V_K$ with a $\pm V_1$ peak-to-peak value at the output of op-amp $OA_5$.

The square wave $V_N$ is converted into triangular wave $V_{T2}$ with a $\pm V_p$ peak-to-peak value by the integrator formed by resistor $R_4$, capacitor $C_2$ and op-amp $OA_2$. For one transition the integrator output is given as

$$V_{T2} = -\frac{1}{R_4 C_2}\int (-V_3)dt = \frac{V_3}{R_4 C_2} t \tag{10.24}$$

From the waveforms shown in Fig. 10.14 the equation (10.24), and the fact that at $t = T/2$, $V_{T2} = 2V_p$ we get

$$2V_P = \frac{V_3}{R_4 C_2}\frac{T}{2}$$

$$V_P = \frac{V_3}{4R_4 C_2} K \frac{V_2}{V_1}$$

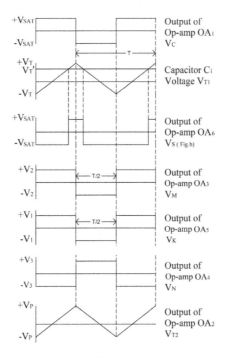

**FIGURE 10.14** Associated waveforms of Fig. 10.13

Let us assume $K = 4R4C2$

$$V_P = \frac{V_2 V_3}{V_1} \qquad (10.25)$$

(i) In Figs. 10.13(a) and (b), the triangular wave $V_{T2}$ is given to the op-amp $OA_6$, diode $D_1$, and capacitor $C_3$ peak detector to get its peak value $V_P$. Hence the output of peak detector will be $V_O = V_P$.

(ii) In Figs. 10.13(c) and (d), the peak value $V_P$ of the triangular wave $V_{T2}$ is obtained by the sample and hold circuit realized by op-amp $OA_6$, switch $S_4$ and capacitor $C_3$. The sampling pulse $V_S$ is generated by comparing first triangular wave $V_{T1}$ with a voltage $V_T'$ that is slightly $V_T$. The sampled output $V_O = V_P$

$$V_O = \frac{V_2 V_3}{V_1} \qquad (10.26)$$

Design exercise:

The switches in Figs. 10.13(a) and (d) are to be replaced with transistorized switches, FET switches and MOSFET switches. In each, (i) draw the circuit diagrams, (ii) explain their working operations, (iii) draw the waveforms at appropriate places, and (iv) deduce the expression for their output voltages.

# 11 Time-Division MCD without Reference

A pulse train whose maximum value is proportional to one voltage ($V_3$) is generated. If the width of this pulse train is made proportional to one voltage ($V_2$) and inversely proportional to another voltage ($V_1$), then the average value of pulse train is proportional to $\dfrac{V_2 V_3}{V_1}$. This is called a time-division multiplier-cum-divider (MCD). There are six types of time-division MCDs.

- (i) Sawtooth wave-referenced MCD – multiplexing
- (ii) Triangular wave-referenced MCD – multiplexing
- (iii) MCD using no reference – multiplexing
- (iv) Sawtooth wave-referenced MCD – switching
- (v) Triangular wave-referenced MCD – switching
- (vi) MCD using no reference – switching

The MCD using no reference, both multiplexing and switching types, is discussed in this chapter.

## 11.1 TIME-DIVISION MCD TYPE I – MULTIPLEXING

The MCDs using a time-division principle without using any reference clock are shown in Fig. 11.1 and their associated waveforms are shown in Fig 11.2. Let's initially assume the comparator $OA_2$ output is LOW. The multiplexer $M_1$ connects ($-V_1$) to the differential integrator composed by resistor $R_1$, capacitor $C_1$ and op-amp $OA_1$ ('ax' is connected to 'a'). The output of differential integrator will be

$$V_{T1} = \frac{1}{R_1 C_1} \int (V_2 + V_1) dt - V_T$$

$$V_{T1} = \frac{(V_1 + V_2)}{R_1 C_1} t - V_T \qquad (11.1)$$

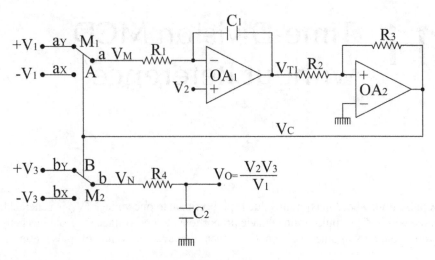

**FIGURE 11.1(A)**   Time-division MCD without reference clock

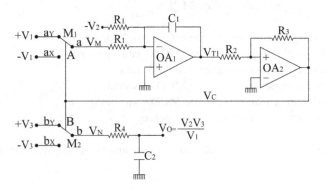

**FIGURE 11.1(B)**   Equivalent circuit of Fig. 11.1(a)

The output of the differential integrator rises towards positive saturation and when it reaches the voltage level of '+$V_T$,' the comparator $OA_2$ output becomes HIGH. The multiplexer $M_1$ connects (+$V_1$) to the differential integrator composed by resistor $R_1$, capacitor $C_1$, and op-amp $OA_1$ ('ay' is connected to 'a'). Now the output of differential integrator will be

$$V_{T1} = \frac{1}{R_1 C_1} \int (V_2 - V_1) dt + V_T$$

$$V_{T1} = -\frac{(V_1 - V_2)}{R_1 C_1} t + V_T \qquad (11.2)$$

# Time-Division MCD without Reference

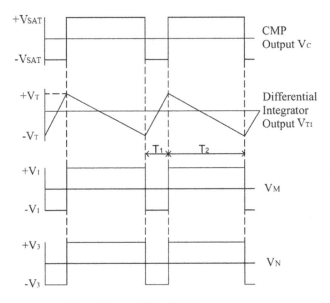

**FIGURE 11.2** Associated waveforms of Fig. 11.1

The output of the differential integrator reverses towards negative saturation and when it reaches the voltage level $(-V_T)$, the comparator $OA_2$ output becomes LOW and the cycle therefore repeats, to give an asymmetrical rectangular wave $V_C$ at the output of comparator $OA_2$.

$$V_T = V_{SAT} \frac{R_2}{R_3} \tag{11.3}$$

From the waveforms shown in Fig. 11.2, it is observed that

$$T_1 = \frac{V_1 - V_2}{2V_1} T, \; T_2 = \frac{V_1 + V_2}{2V_1} T, \; T = T_1 + T_2 \tag{11.4}$$

The asymmetrical rectangular wave $V_C$ controls another multiplexer $M_2$. The multiplexer $M_2$ selects $(+V_3)$ during the ON time $T_2$ ('by' is connected to 'b') and $(-V_3)$ during the OFF time $T_1$ of the rectangular wave $V_C$ ('bx' is connected to 'b'). Another rectangular wave $V_N$ is generated at the multiplexer $M_2$ output. The $R_4C_2$ low pass filter gives the average value of this pulse train $V_N$ and is given as

$$V_O = \frac{1}{T} \left[ \int_0^{T_2} V_3 \, dt + \int_{T_2}^{T_1+T_2} (-V_3) \, dt \right]$$

$$V_O = \frac{V_3(T_2 - T_1)}{T} \tag{11.5}$$

Equation (11.4) in (11.5) gives

$$V_O = \frac{V_2 V_3}{V_1} \tag{11.6}$$

Design exercise:

1. Replace multiplexers $M_1$ and $M_2$ in Figs. 11.1(a) and (b) with a transistor multiplexer, FET multiplexer, and MOSFET multiplexer. In each, (i) draw circuit diagrams, (ii) draw waveforms at appropriate places, (iii) explain the working operation, and (iv) deduce the expression for their output voltages.
2. The inputs of the multiplexer $M_2$ in Figs. 11.1(a) and (b) are interchanged. (i) Draw waveforms at appropriate places, (ii) explain working operation, and (iii) deduce the expression for their output voltages.

## 11.2 TIME-DIVISION MCD TYPE II – MULTIPLEXING

The MCDs using the time-division principle without using any reference clock are shown in Fig. 11.3 and their associated waveforms are shown in Fig 11.4.

Let's initially assume the comparator $OA_2$ output is LOW. The multiplexer $M_1$ connects $(-V_3)$ to the inverting terminal of a differential integrator ('ax' is connected 'a'). The output of the differential integrator will be

$$V_{T1} = \frac{1}{R_1 C_1} \int (V_O + V_3) dt - V_T$$

$$V_{T1} = \frac{(V_O + V_3)}{R_1 C_1} t - V_T \tag{11.7}$$

The output of the differential integrator is rising towards positive saturation and when it reaches the voltage level of '$+V_T$,' the comparator $OA_2$ output becomes HIGH. $+V_{SAT}$ is given to the inverting terminal of the differential integrator. The switch $S_1$ connects $(+V_3)$ to the inverting terminal of the differential integrator. Now the output of differential integrator will be

$$V_{T1} = \frac{1}{R_1 C_1} \int (V_O - V_3) dt + V_T$$

$$V_{T1} = -\frac{(V_3 - V_O)}{R_1 C_1} t + V_T \tag{11.8}$$

# Time-Division MCD without Reference

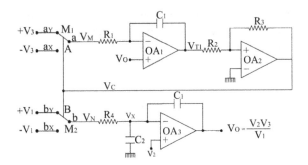

**FIGURE 11.3(A)** MCD without reference clock – II

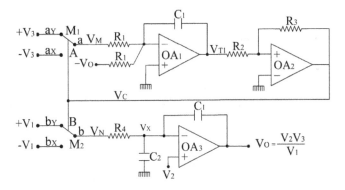

**FIGURE 11.3(B)** Equivalent circuit of Fig. 11.3(a)

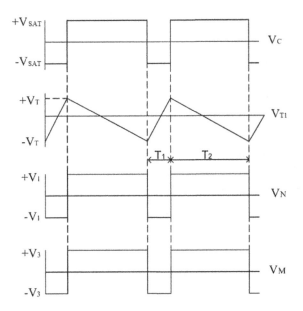

**FIGURE 11.4** Associated waveforms of Fig. 11.3

The output of the differential integrator reverses towards negative saturation and when it reaches the voltage level $(-V_T)$, the comparator $OA_2$ output becomes LOW and the cycle therefore repeats, to give an asymmetrical rectangular wave $V_C$ at the output of comparator $OA_2$.

$$V_T = \frac{R_2}{R_3} V_{SAT} \tag{11.9}$$

From the waveforms shown in Fig. 11.4, it is observed that

$$T_1 = \frac{V_3 - V_O}{2V_3} T, \ T_2 = \frac{V_3 + V_O}{2V_3} T, \ T = T_1 + T_2 \tag{11.10}$$

The asymmetrical rectangular wave $V_C$ controls another multiplexer $M_2$. The switch $S_2$ selects $(+V_1)$ during the ON time $T_2$ ('by' is connected to 'b') and $(-V_1)$ during the OFF time $T_1$ of the rectangular wave $V_C$ ('bx' is connected to 'b'). Another rectangular wave $V_N$ with $\pm V_1$ as the peak-to-peak value is generated at the multiplexer $M_2$ output. The $R_4C_2$ low pass filter gives the average value of this pulse train $V_N$ and is given as

$$V_X = \frac{1}{T}\left[\int_O^{T_2} V_1 \, dt + \int_{T_2}^{T_1+T_2} (-V_1) \, dt\right]$$

$$V_X = \frac{V_1(T_2 - T_1)}{T} \tag{11.11}$$

Equations (11.10) in (11.11) gives

$$V_X = \frac{V_O V_1}{V_3} \tag{11.12}$$

The op-amp OA3 has a negative closed loop configuration and a positive DC voltage is ensured in the feedback loop. Hence its non-inverting terminal voltage is equal to its inverting terminal voltage, i.e.

$$V_2 = V_X \tag{11.13}$$

From equations (11.12) and (11.13)

$$V_O = \frac{V_2 V_3}{V_1} \tag{11.14}$$

Design exercise:

Replace multiplexers $M_1$ and $M_2$ in Figs. 11.3(a) and (b) with a transistor multiplexer, FET multiplexer and MOSFET multiplexer. In each, (i) draw circuit diagrams, (ii) draw waveforms at appropriate places, (iii) explain the working operation, and (iv) deduce the expression for their output voltages

## 11.3 TIME-DIVISION MCD TYPE I – SWITCHING

The MCDs using the time-division principle without using any reference clock are shown in Fig 11.5 and their associated waveforms in Fig 11.6. Figs. 11.5(a) and (b) show a series-switching MCD and Figs. 11.5(c) and (d) show a parallel-switching MCD. Let's initially assume the comparator $OA_2$ output is LOW.

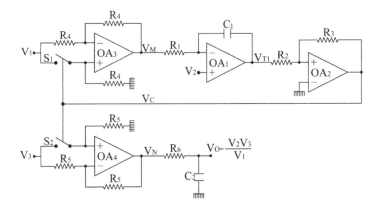

**FIGURE 11.5(A)**  Series-switching time-division MCD without reference clock

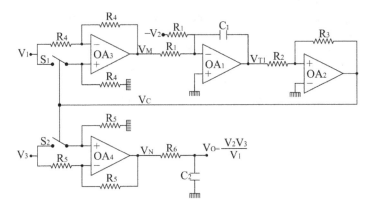

**FIGURE 11.5(B)**  Equivalent circuit of Fig. 11.5(a)

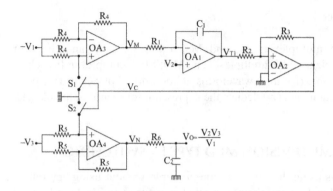

**FIGURE 11.5(C)** Shunt-switching time-division MCD

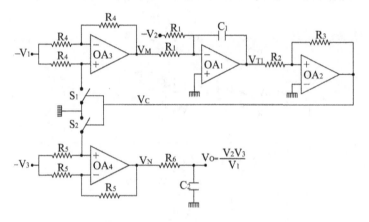

**FIGURE 11.5(D)** Equivalent circuit of Fig. 11.5(c)

(i) In Figs 11.5(a) and (b), the switch $S_1$ is open, op-amp $OA_3$ along with resistors $R_4$ will work as an inverting amplifier and $(-V_1)$ will be at its output $(V_M = -V_1)$. The switch $S_2$ is open, op-amp $OA_4$ along with resistors $R_5$ will work as an inverting amplifier, and $(-V_3)$ will be at its output $(V_N = -V_3)$.

(ii) In Figs. 11.5(c) and (d), the switch $S_1$ is open, op-amp $OA_3$ along with resistors $R_4$ will work as a non-inverting amplifier and $(-V_1)$ will be at its output $(V_M = -V_1)$. The switch $S_2$ is open, op-amp $OA_4$ along with resistors $R_5$ will work as a non-inverting amplifier and $(-V_3)$ will be at its output $(V_N = -V_3)$

The output of differential integrator will be

$$V_{T1} = \frac{1}{R_1 C_1} \int (V_2 + V_1) dt - V_T$$

# Time-Division MCD without Reference

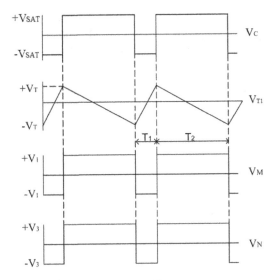

**FIGURE 11.6** Associated waveforms of Fig. 11.5

$$V_{T1} = \frac{(V_2 + V_1)}{R_1 C_1} t - V_T \qquad (11.15)$$

The output of the differential integrator is rising towards positive saturation and when it reaches the voltage level of '$+V_T$,' the comparator $OA_2$ output becomes HIGH.

(i) In Figs. 11.5(a) and (b), the switch $S_1$ is closed, op-amp $OA_3$ along with resistors $R_4$ will work as a non-inverting amplifier and $(+V_1)$ will be at its output $(V_M = +V_1)$. The switch $S_2$ is closed, op-amp $OA_4$ along with resistors $R_5$ will work as a non-inverting amplifier and $(+V_3)$ will be at its output $(V_N = +V_3)$.

(ii) In Figs 11.5(c) and (d), the switch $S_1$ is closed, op-amp $OA_3$ along with resistors $R_4$ will work as an inverting amplifier and $(+V_1)$ will be at its output $(V_M = +V_1)$. The switch $S_2$ is closed, op-amp $OA_4$ along with resistors $R_5$ will work as inverting amplifier and $(+V_3)$ will be at its output $(V_N = +V_3)$

Now the output of differential integrator will be

$$V_{T1} = \frac{1}{R_1 C_1} \int (V_2 - V_1) dt + V_T$$

$$V_{T1} = -\frac{(V_1 - V_2)}{R_1 C_1} t + V_T \qquad (11.16)$$

The output of differential integrator reverses towards negative saturation and when it reaches the voltage level $(-V_T)$, the comparator $OA_2$ output becomes LOW and the cycle therefore repeats, to give an asymmetrical rectangular wave $V_C$ at the output of comparator $OA_2$.

$$V_T = \frac{R_2}{R_3} V_{SAT} \tag{11.17}$$

From the waveforms shown in Fig. 11.6, it is observed that

$$T_1 = \frac{V_1 - V_2}{2V_1} T, \quad T_2 = \frac{V_1 + V_2}{2V_1} T, \quad T = T_1 + T_2 \tag{11.16}$$

Another rectangular wave $V_N$ is generated at the output of op-amp $OA_4$. The $R_6 C_2$ low pass filter gives the average value of this pulse train $V_N$ and is given as

$$V_O = \frac{1}{T}\left[\int_0^{T_2} V_3\, dt + \int_{T_2}^{T_1+T_2} (-V_3)\, dt\right]$$

$$V_O = \frac{V_3(T_2 - T_1)}{T} \tag{11.19}$$

Equation (11.18) in (11.19) gives

$$V_O = \frac{V_2 V_3}{V_1} \tag{11.20}$$

Design exercise:

1. The switches $S_1$ and $S_2$ in Figs. 11.5(a) to (d) can be replaced with transistorized switches as shown in Figs. 11.7(a) to (d) respectively. (i) Explain the working operation of the multipliers shown in Figs. 11.3(a) and (b), (ii) draw the waveforms at appropriate places, and (iii) deduce the expression for their output voltages.
2. The switches $S_1$ and $S_2$ in Figs. 11.5(a) to (d) are to be replaced with FET switches and MOSFET switches. In each, (i) draw the circuit diagrams, (ii) explain their working operations, (iii) draw waveforms at appropriate places, and (iv) deduce the expression for their output voltages.

# Time-Division MCD without Reference

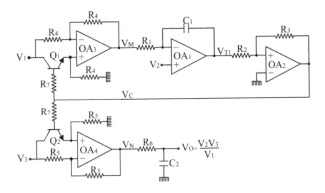

**FIGURE 11.7(A)** Series-switching time-division MCD with no reference

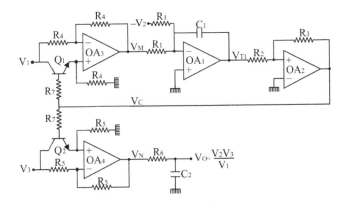

**FIGURE 11.7(B)** Equivalent circuit of Fig. 11.7(a)

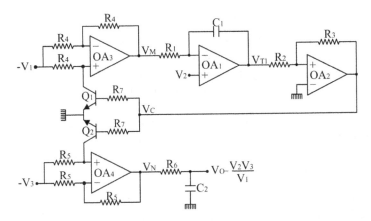

**FIGURE 11.7(C)** Shunt-switching time-division MCD with no reference

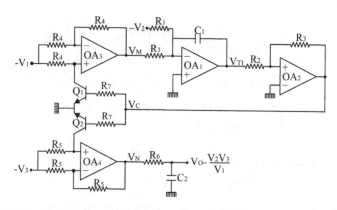

**FIGURE 11.7(D)** Equivalent circuit of Fig. 11.7(c)

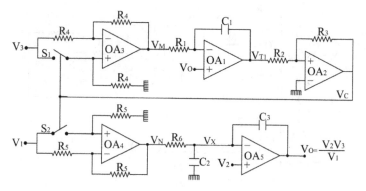

**FIGURE 11.8(A)** Series-switching MCD without reference clock – II

## 11.4 TIME-DIVISION MCD TYPE II – SWITCHING

The MCDs using the time-division principle without using any reference clock are shown in Fig 11.8 and their associated waveforms are shown in Fig. 11.9. Figs. 11.8(a) and (b) show series-switching MCDs and Figs. 11.8(c) and (d) show shunt-switching MCDs. Let's initially assume the comparator $OA_2$ output is LOW.

(i) In Figs. 11.8(a) and (b), the switch $S_1$ is opened, op-amp $OA_3$ along with resistors $R_4$ will work as inverting amplifier and $(-V_3)$ will be at its output. The switch $S_2$ is opened, op-amp $OA_4$ along with resistors $R_5$ will work as inverting amplifier and $(-V_1)$ will be at its output.

(ii) In Figs. 11.8(c) and (d), the switch $S_1$ is opened, op-amp $OA_3$ along with resistors $R_4$ will work as non-inverting amplifier and $(-V_3)$ will be at its output. The switch $S_2$ is opened, op-amp $OA_4$ along with resistors $R_5$ will work as non-inverting amplifier, and $(-V_1)$ will be at its output.

# Time-Division MCD without Reference

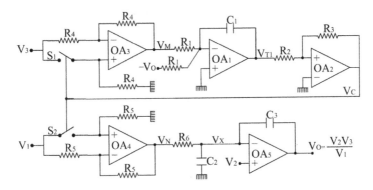

**FIGURE 11.8(B)** Equivalent circuit of Fig. 11.8(a)

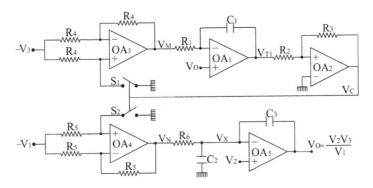

**FIGURE 11.8(C)** Shunt-switching MCD

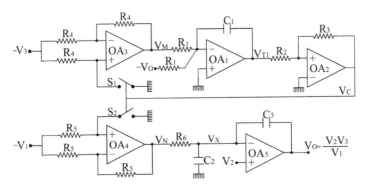

**FIGURE 11.8(D)** Equivalent circuit of Fig. 11.8(c)

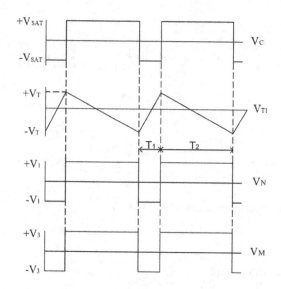

**FIGURE 11.9** Associated waveforms of Fig. 11.8

The output of the differential integrator will be

$$V_{T1} = \frac{1}{R_1 C_1}\int (V_O + V_3)dt - V_T$$

$$V_{T1} = \frac{(V_3 + V_O)}{R_1 C_1} t - V_T \qquad (11.21)$$

The output of the differential integrator is rising towards positive saturation and when it reaches the voltage level of '$+V_T$,' the comparator $OA_2$ output becomes HIGH.

(i) In Figs 11.8(a) and (b), the switch $S_1$ is closed, op-amp $OA_3$ along with resistors $R_4$ will work as a non-inverting amplifier and $(+V_3)$ will be at its output. The switch $S_2$ is closed, op-amp $OA_4$ along with resistors $R_5$ will work as a non-inverting amplifier, and $(+V_1)$ will be at its output

(ii) In Figs 11.8(c) and (d), the switch $S_1$ is closed, op-amp $OA_3$ along with resistors $R_4$ will work as inverting amplifier and $(+V_3)$ will be at its output. The switch $S_2$ is closed, op-amp $OA_4$ along with resistors $R_5$ will work as inverting amplifier and $(+V_1)$ will be at its output.

Now the output of differential integrator will be

$$V_S = \frac{1}{R_1 C_1}\int (V_O - V_3)dt + V_T$$

# Time-Division MCD without Reference

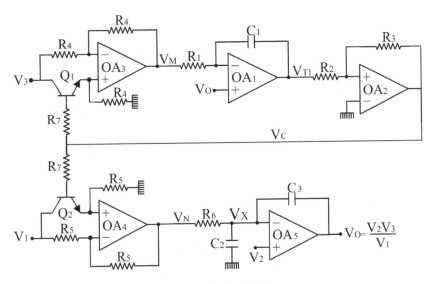

**FIGURE 11.10(A)** Series-switching time-division MCD

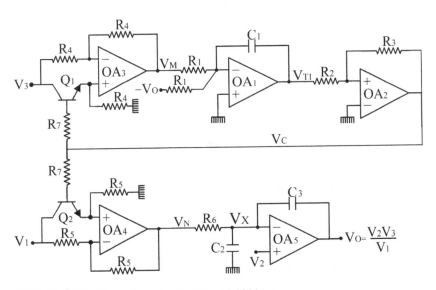

**FIGURE 11.10(B)** Equivalent circuit of Fig. 11.10(a)

$$V_S = -\frac{(V_3 - V_O)}{R_1 C_1} t + V_T \qquad (11.21)$$

The output of differential integrator reverses towards negative saturation and when it reaches the voltage level $(-V_T)$, the comparator $OA_2$ output becomes LOW and the

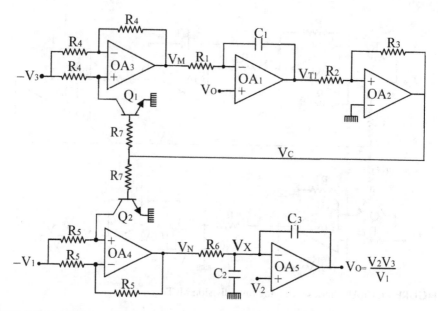

**FIGURE 11.10(C)**  Shunt-switching time-division MCD with no reference

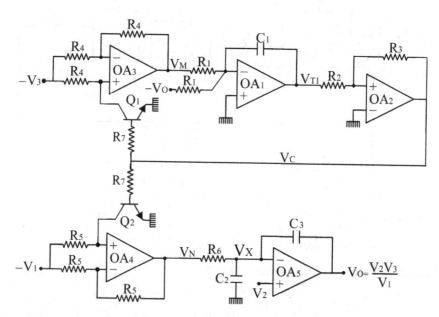

**FIGURE 11.10(D)**  Equivalent circuit of Fig. 11.10(c)

## Time-Division MCD without Reference

cycle therefore repeats, to give an asymmetrical rectangular wave $V_C$ at the output of comparator CMP.

$$V_T = \frac{R_2}{R_3} V_{SAT} \tag{11.23}$$

From the waveforms shown in Fig.11.9, it is observed that

$$T_1 = \frac{V_3 - V_O}{2V_3} T, \; T_2 = \frac{V_3 + V_O}{2V_3} T, \; T = T_1 + T_2 \tag{11.24}$$

Another rectangular wave $V_N$ with $\pm V_1$ peak-to-peak values is generated at the output of op-amp $OA_4$. The $R_6C_2$ low pass filter gives the average value of this pulse train $V_N$ and is given as

$$V_X = \frac{1}{T}\left[\int_O^{T_2} V_1 \, dt + \int_{T_2}^{T_1+T_2} (-V_1) \, dt\right]$$

$$V_X = \frac{V_1(T_2 - T_1)}{T} \tag{11.25}$$

Equations (11.24) in (11.25) gives

$$V_X = \frac{V_1 V_O}{V_3} \tag{11.26}$$

The op-amp $OA_5$ is at negative closed loop feedback configuration and a positive DC voltage is ensured in the feedback. Its non-inverting terminal voltage will be equal to its inverting terminal voltage. Hence

$$V_X = V_2 \tag{11.27}$$

From equations (11.26) and (11.27)

$$V_O = \frac{V_2 V_3}{V_1} \tag{11.28}$$

Design exercise:

1. The switches $S_1$ and $S_2$ in Figs. 11.8(a) to (d) can be replaced with transistorized switches as shown in Figs. 11.10(a) to (d) respectively. (i) Explain the

working operation of the multipliers shown in Figs. 11.3(a) and (b), (ii) draw the waveforms at appropriate places, and (iii) deduce the expression for their output voltages.
2. The switches $S_1$ and $S_2$ in Figs. 11.8(a) to (d) are to be replaced with FET switches and MOSFET switches. In each, (i) draw the circuit diagrams, (ii) explain their working operations, (iii) draw the waveforms at appropriate places, and (iv) deduce the expression for their output voltages.

# 12 Pulse Position-Responding MCDs

Pulse position peak-detecting MCDs, pulse position-sampled MCDs and pulse-width integrated peak-sampling MCDs are called pulse position-responding MCDs and discussed in this chapter.

A sawtooth wave of period T whose peak value is proportional to one input voltage ($V_3$) is connected to a peak detector during a time $\delta_T$, which is proportional to another input voltage ($V_2$). The time period T of this sawtooth wave is proportional to one input voltage ($V_1$). The peak detector output is proportional to $\dfrac{V_2 V_3}{V_1}$. $\delta_T$ is obtained from T and $\delta_T < T$. This is called a pulse position peak-detecting MCD.

A sawtooth wave of period T whose (i) time period T is proportional to one input voltage ($V_1$) and (ii) peak value is proportional to one input voltage ($V_3$) is sampled by a sampling pulse whose position over the period T is proportional to another input voltage ($V_2$). The sampled output is proportional to $\dfrac{V_2 V_3}{V_1}$. This is called a pulse position-sampled MCD.

## 12.1 PULSE POSITION PEAK-DETECTING MCDs – MULTIPLEXING

The circuit diagrams of multiplexing type pulse position peak-detecting MCDs are shown in Fig. 12.1 and their associated waveforms are shown in Fig. 12.2.

In Fig. 12.1(a), as discussed in Section 5.1 [Fig.5.1(a)], the op-amp $OA_1$, transistor $Q_1$, multiplexer $M_1$ constitutes a sawtooth wave generator.

In Fig. 12.1(b), as discussed in Section 5.1 [Fig. 5.1(b)], the op-amps $OA_1$ and $OA_2$ and multiplexer $M_1$ constitutes a sawtooth wave generator. In both circuits, the time period of the sawtooth wave $V_{S1}$ is given as

$$T = \frac{V_R}{V_1} R_1 C_1 \qquad (12.1)$$

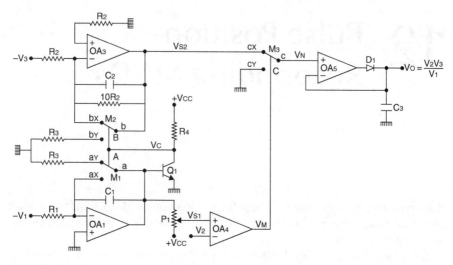

**FIGURE 12.1(A)** Pulse position peak-detecting MCD – type I

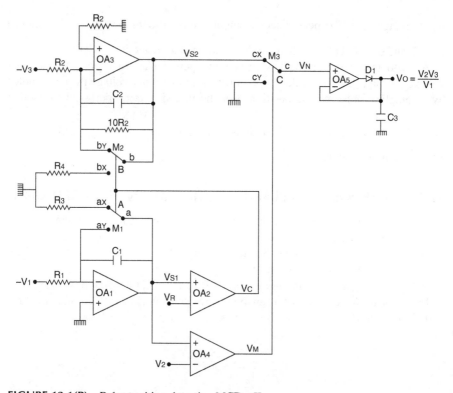

**FIGURE 12.1(B)** Pulse position-detecting MCD – II

# Pulse Position-Responding MCDs

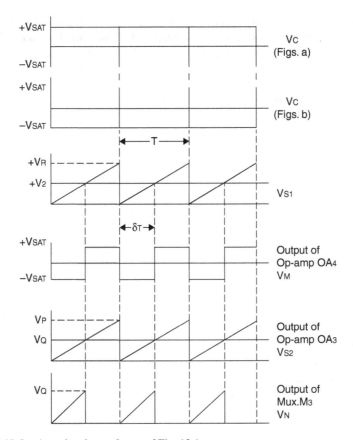

**FIGURE 12.2** Associated waveforms of Fig. 12.1

The comparator $OA_4$ compares this sawtooth wave $V_{S1}$ with an input voltage $V_2$ and produces a rectangular wave $V_M$. The OFF time $\delta_T$ of this rectangular $V_M$ is given as

$$\delta_T = \frac{V_2}{V_R}T \qquad (12.2)$$

The short-pulse $V_C$ in the sawtooth wave generator is also given to multiplexer $M_2$, which constitutes a controlled integrator along with op-amp $OA_3$, resistor $R_2$, and capacitor $C_2$.

(i) In Fig. 12.1(a), during the HIGH value of $V_C$, the multiplexer $M_2$ connects 'by' to 'b,' and another integrator is formed by op-amp $OA_3$, resistor $R_2$, and capacitor $C_2$. During the LOW value of $V_C$, the multiplexer $M_2$ connects 'bx' to 'b,' and capacitor $C_2$ is short-circuited so that the integrator $OA_3$ output becomes zero.

(ii) In Fig. 12.1(b), during the HIGH value of $V_C$, the multiplexer $M_2$ connects 'by' to 'b,' and capacitor $C_2$ is short-circuited so that integrator $OA_3$ output becomes zero. During the LOW value of $V_C$, the multiplexer $M_2$ connects 'bx' to 'b,' and another integrator is formed by op-amp $OA_3$, resistor $R_2$, and capacitor $C_2$.

The integrator $OA_3$ output is given as

$$V_{S2} = -\frac{1}{R_2 C_2} \int -V_3 dt = \frac{V_3}{R_2 C_2} t \qquad (12.3)$$

Another sawtooth wave $V_{S2}$ with a peak value of $V_P$ is generated at the output of the integrator $OA_3$. From the waveforms shown in Fig 12.2, from equation (12.3) and the fact that at $t = T$, $V_{S2} = V_P$ we get

$$V_P = \frac{V_3}{R_2 C_2} T \qquad (12.4)$$

The rectangular pulse $V_M$ controls multiplexer $M_3$. During the LOW of $V_M$, the multiplexer $M_3$ connects 'cx' to 'c' and the sawtooth wave $V_{S2}$ is connected to the multiplexer $M_3$ output. During the HIGH of $V_M$, the multiplexer $M_3$ connects 'cy' to 'c' and zero volts is connected to the multiplexer $M_3$. A semi-sawtooth wave $V_N$ with the peak value $V_Q$ is generated at the output of multiplexer $M_3$. The peak detector realized by op-amp $OA_5$, diode $D_1$, and capacitor $C_3$ gives this peak value $V_Q$ at its output, i.e. $V_O = V_Q$.
The peak value $V_Q$ is given as

$$V_Q = \frac{V_P}{T} \delta_T \qquad (12.5)$$

From equations (12.2), (12.4) and (12.5)

$$V_O = \frac{V_3}{R_2 C_2} \frac{V_2}{V_R} T \qquad (12.6)$$

Equation (12.1) in (12.6) gives

$$V_O = \frac{V_2 V_3}{V_1} \frac{R_1 C_1}{R_2 C_2} \qquad (12.7)$$

Let $R_1 = R_2$, $C_1 = C_2$

$$V_O = \frac{V_2 V_3}{V_1} \qquad (12.8)$$

Pulse Position-Responding MCDs

Design exercise:

1. In the MCD circuits shown in Figs. 12.1(a) and (b), if the polarity of input voltage $V_3$ and direction of diode $D_1$ are reversed, (i) draw waveforms at appropriate places and (ii) deduce the expression for their output voltages
2. In the MCD circuits shown in Figs. 12.1(a) and (b), the input terminals of comparator $OA_4$ and multiplexer $M_3$ are interchanged. (i) Draw the circuit diagrams, (ii) explain their working operations, (iii) draw waveforms at appropriate places, and (iv) Deduce expression for their output voltages.

## 12.2 PULSE POSITION PEAK-DETECTING MCDs – SWITCHING

The circuit diagram of a switching type pulse position peak-detecting MCD is shown in Fig. 12.3 and its associated waveforms are shown in Fig. 12.4. Let's initially assume the op-amp $OA_2$ output is LOW, the switch $S_1$ opens, and an integrator formed by resistor $R_1$, capacitor $C_1$, and op-amp $OA_1$ integrates the input voltage $(-V_1)$. The integrator output is given as

$$V_{S1} = -\frac{1}{R_1 C_1}\int -V_1 dt = \frac{V_1}{R_1 C_1} t \qquad (12.9)$$

A positive-going ramp $V_{S1}$ is generated at the output of op-amp $OA_1$. When the output of $OA_1$ reaches the voltage level of $+V_R$, the comparator $OA_2$ output becomes HIGH.

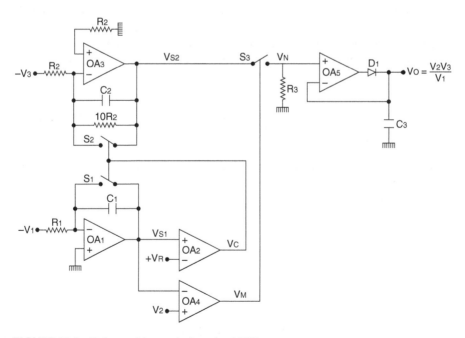

**FIGURE 12.3** Pulse position peak-detecting MCD

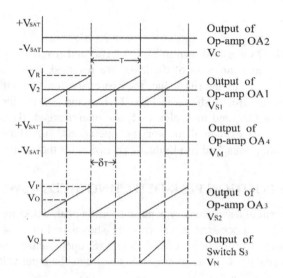

**FIGURE 12.4** Associated waveforms of Fig. 12.3

The switch $S_1$ is closed and hence the capacitor $C_1$ is shorted so that the op-amp $OA_1$ output becomes ZERO. Then op-amp $OA_2$ output goes to LOW, the switch $S_1$ opens, and the integrator composed by $R_1$, $C_1$, and op-amp $OA_1$ integrates the input voltage $(-V_1)$ and the cycle therefore repeats to provide (i) a sawtooth wave of peak value $V_R$ at the output of op-amp $OA_1$ and (ii) a short-pulse waveform $V_C$ at the output of comparator $OA_2$. The short pulse $V_C$ also controls switch $S_2$. During the short HIGH time of $V_C$, switch $S_2$ is closed, and the capacitor $C_2$ is short-circuited so that the op-amp $OA_3$ output is zero volts. During LOW time of $V_C$, switch $S_2$ opens, and the integrator formed by resistor $R_2$, capacitor $C_2$, and op-amp $OA_3$ integrates its input voltage $(-V_3)$ and its output is given as

$$V_{S2} = -\frac{1}{R_2 C_2}\int -V_3 dt = \frac{V_3}{R_2 C_2}t \qquad (12.10)$$

Another sawtooth waveform $V_{S2}$ with peak value $V_P$ is generated at the output of op-amp $OA_3$. From the waveforms shown in Fig. 12.4 from equations (12.9), (12.10) and the fact that at t = T,

$$V_{S1} = V_R, V_{S2} = V_P.$$

$$V_R = \frac{V_1}{R_1 C_1}T, \; T = \frac{V_R}{V_1}R_1 C_1$$

$$V_P = \frac{V_3}{R_2 C_2}T, \; V_P = \frac{V_3}{R_2 C_2}\frac{V_R}{V_1}R_1 C_1$$

# Pulse Position-Responding MCDs

Let us assume $R_1 = R_2$, $C_1 = C_2$, then

$$V_P = V_3 \frac{V_R}{V_1} \quad (12.11)$$

The sawtooth wave $V_{S1}$ is compared with second input voltage $V_2$ by the comparator $OA_4$ and a rectangular wave $V_M$ is generated at the output of comparator $OA_4$.
The ON time of this rectangular waveform $V_M$ is given as

$$\delta_T = \frac{V_2}{V_R} T \quad (12.12)$$

The rectangular pulse $V_M$ controls the switch $S_3$. During the HIGH of $V_M$, the switch $S_3$ is closed and the sawtooth wave $V_{S2}$ is connected to the peak detector realized by op-amp $OA_5$, diode $D_1$, and capacitor $C_3$. During the LOW of $V_M$, the switch $S_3$ is open and zero volts exists on the peak detector. A semi-sawtooth wave $V_N$ with peak value $V_Q$ is generated at the output of switch $S_3$. The peak detector realized by diode $D_1$ and capacitor $C_3$ gives this peak value $V_Q$ at its output. Hence $V_O = V_Q$.

The peak value $V_Q$ is given as

$$V_Q = \frac{V_P}{T} \delta_T \quad (12.13)$$

$$V_Q = V_O$$

Equations (12.11) and (12.12) in (12.13) give

$$V_O = \frac{V_2 V_3}{V_1} \quad (12.14)$$

Design exercise:

1. The switches $S_1$–$S_3$ in Fig. 12.3 are to be replaced with transistorized switches as shown in Fig. 12.5. (i) Explain its working operation, (ii) draw the waveforms at various points, and (iii) deduce the expression for their output voltages.
2. The switches $S_1$–$S_3$ in Fig. 12.3 are to be replaced with FET switches and MOSFET switches. In each, (i) draw the circuit diagrams, (ii) explain their working operations, (iii) draw waveforms at appropriate places, and (iv) deduce the expression for their output voltages.

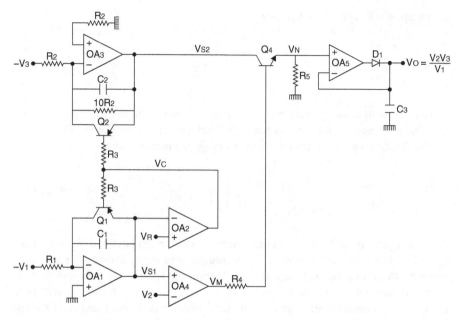

**FIGURE 12.5** Pulse position peak-detecting MCD

## 12.3 PULSE POSITION PEAK-SAMPLING MCD – MULTIPLEXING

The circuit diagram of pulse position sampling MCD is shown in Fig. 12.6 and its associated waveforms are shown in Fig. 12.7. As discussed in Section 5.1, op-amp $OA_1$ along with transistor $Q_1$ and multiplexer $M_1$ constitute a sawtooth wave generator. The time period of the generated sawtooth wave is given as

$$T = \frac{V_R}{V_1} R_1 C_1 \qquad (12.15)$$

Where $V_R$ is the $V_{BE}$ value of transistor $Q_1$.

This sawtooth wave $V_{S1}$ is compared with the second input voltage $V_2$ by the comparator $OA_2$ to get a rectangular pulse $V_M$. The OFF time of this rectangular wave $V_M$ is given as

$$\delta_T = \frac{V_2}{V_R} T \qquad (12.16)$$

The sawtooth wave $V_{S1}$ is also compared with a voltage $V_2$', which is slightly less than $V_2$, by the comparator $OA_3$ to get another rectangular pulse $V_N$. The two rectangular pulses $V_M$ and $V_N$ are given to an Ex-OR gate. The output of the Ex-OR gate is a short-pulse $V_S$, which is acting as a sampling pulse to the sample and hold circuit realized by multiplexer $M_3$ and capacitor $C_3$.

## Pulse Position-Responding MCDs

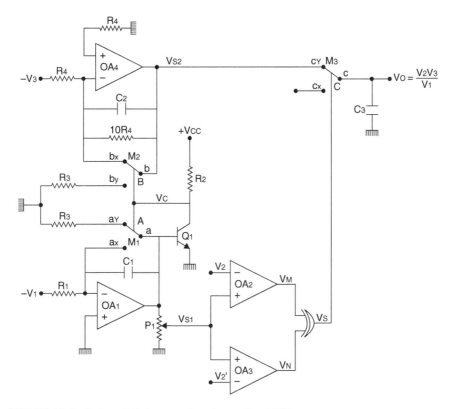

**FIGURE 12.6** Pulse-width integrated peak-sampling MCD

The short-pulse $V_C$ in the sawtooth wave generator is also given to multiplexer $M_2$, which constitutes a controlled integrator along with the op-amp $OA_4$, resistor $R_4$, and capacitor $C_2$.

During the ON time of $V_C$, the multiplexer $M_2$ selects 'by' to 'b' and the integrator formed by resistors $R_4$, capacitor $C_2$, and op-amp $OA_4$ integrates its input voltage $(-V_3)$ and is given as

$$V_{S2} = -\frac{1}{R_4 C_2}\int -V_3 dt = \frac{V_3}{R_4 C_2} t \qquad (12.17)$$

During the OFF time of $V_C$, the multiplexer $M_2$ selects 'bx' to 'b' and the capacitor $C_2$ is short-circuited and hence the output of integrator becomes zero. Another sawtooth wave $V_{S2}$ with a peak value of $V_P$ is generated at the output of op-amp $OA_4$. From the waveforms shown in Fig. 12.7, from equation 12.17 and the fact that at $t = T$, $V_{S2} = V_P$ we get

$$V_P = \frac{V_3}{R_4 C_2} T \qquad (12.18)$$

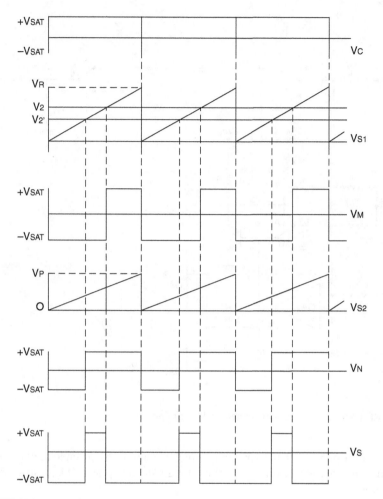

**FIGURE 12.7** Associated waveforms of Fig. 12.6

The sampled output is given as

$$V_O = \frac{V_P}{T} \delta_T \qquad (12.19)$$

Equations (12.15), (12.16) and (12.18) in (12.19) give

$$V_P = \frac{V_2 V_3}{V_1} \frac{R_1 C_1}{R_4 C_2}$$

Pulse Position-Responding MCDs

Let $R_1 = R_4$ and $C_1 = C_2$

$$V_P = \frac{V_2 V_3}{V_1}$$

The peak value $V_P$ of this sawtooth wave $V_{S2}$ is obtained by the sample and hold circuit realized by multiplexer $M_3$ and capacitor $C_3$ with a sampling pulse $V_S$. The sample and hold output $V_O = V_P$.

$$V_O = \frac{V_2 V_3}{V_1} \qquad (12.20)$$

Design exercise:

1. In the multiplier circuits shown in Fig. 12.11, if the polarity of input voltage $V_3$ is reversed, (i) draw waveforms at appropriate places and (ii) deduce the expression for their output voltages.
2. The sawtooth generator in Fig. 12.11 is to be replaced with the other sawtooth wave generators shown in Figs. 5.1(b), (c), and (d). In each, (i) draw circuit diagrams, (ii) draw waveforms at appropriate places, (iii) explain working operation, and (iv) deduce the expression for their output voltages.

## 12.4 PULSE POSITION PEAK-SAMPLING MCD – SWITCHING

The circuit diagram of a pulse-width integrated peak-sampling MCD is shown in Fig. 12.8 and its associated waveforms are shown in Fig. 12.9. As discussed in Section 8.1(a), op-amps $OA_1$ and $OA_2$ along with switch $S_1$ constitutes a sawtooth wave generator. The time period of the generated sawtooth wave is given as

$$T = \frac{V_R}{V_1} R_1 C_1 \qquad (12.21)$$

This sawtooth wave $V_{S1}$ is compared with the second input voltage $V_2$ by the comparator $OA_4$ to get a rectangular pulse $V_M$. The OFF time of this rectangular wave $V_M$ is given as

$$\delta_T = \frac{V_2}{V_R} T \qquad (12.22)$$

The sawtooth wave $V_{S1}$ is also compared with $V_2$', which is slightly less than $V_2$, by the comparator $OA_5$ to get another rectangular pulse $V_N$. The two rectangular pulses $V_M$ and $V_N$ are given to an Ex-OR gate. The output of the Ex-OR gate is a short-pulse

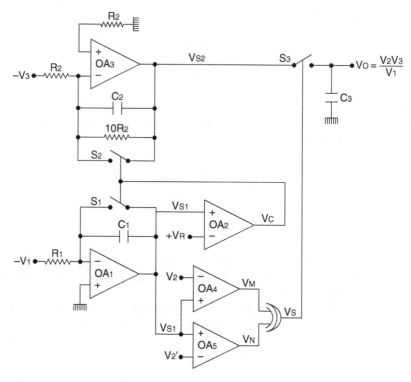

**FIGURE 12.8** Pulse-width integrated peak-sampling MCD

$V_S$, which is acting as a sampling pulse to the sample and hold circuit realized by switch $S_3$ and capacitor $C_3$.

The short pulse $V_C$ in the sawtooth generator is also given to the switch $S_2$, which constitutes a controlled integrator along with the op-amp $OA_3$, resistor $R_2$ and capacitor $C_2$.

During the OFF time of $V_C$, the switch $S_2$ is opened and the integrator formed around op-amp $OA_3$ integrates its input voltage $(-V_3)$ and is given as

$$V_{S2} = -\frac{1}{R_2 C_2}\int -V_3 dt = \frac{V_3}{R_2 C_2}t \qquad (12.23)$$

During the ON time of $V_C$, the switch $S_2$ is closed and the capacitor $C_2$ is short-circuited and hence the output of integrator becomes zero. Another sawtooth wave $V_{S2}$ with peak value of $V_P$ is generated at the output of op-amp $OA_3$. From the waveforms shown in Fig. 12.9, from equation (12.23) and the fact that at $t = T$, $V_{S2} = V_P$

$$V_P = \frac{V_3}{R_2 C_2}T \qquad (12.24)$$

# Pulse Position-Responding MCDs

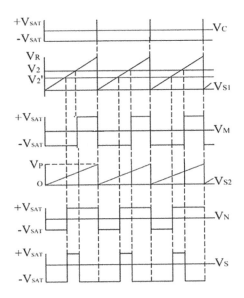

**FIGURE 12.9** Associated waveforms of Fig. 12.8

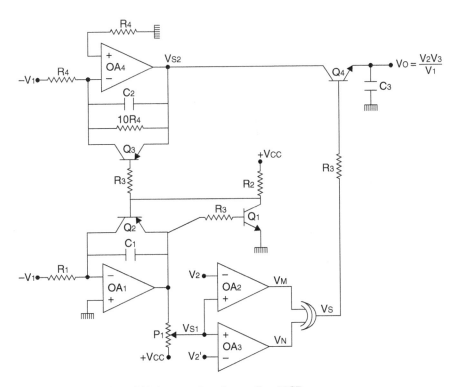

**FIGURE 12.10** Pulse-width integrated peak-sampling MCD

The sampled output is given as

$$V_O = \frac{V_P}{T}\delta_T \tag{12.25}$$

Equations (12.21), (12.22), and (12.24) in (12.25) gives

$$V_P = \frac{V_2 V_3}{V_1}\frac{R_1 C_1}{R_2 C_2}$$

Let $R_1 = R_2$ and $C_1 = C_2$

$$V_P = \frac{V_2 V_3}{V_1}$$

The peak value $V_P$ of this sawtooth wave $V_{s2}$ is obtained by the sample and hold circuit realized by the switch $S_3$ and capacitor $C_3$. The sample and hold output $V_O = V_P$.

$$V_O = \frac{V_2 V_3}{V_1} \tag{12.26}$$

Design exercise:

1. The switches $S_1$–$S_3$ in Fig. 12.8 are replaced with transistorized switches and shown in Figs. 12.10. (i) Explain its working operation, (ii) draw waveforms at various points, and (iii) deduce the expression for their output voltages.
2. The switches $S_1$–$S_3$ in Figs. 12.8 are to be replaced with FET switches and MOSFET switches. In each, (i) draw the circuit diagrams, (ii) explain its working operations, (iii) draw waveforms at appropriate places, and (iv) deduce the expression for their output voltages.
3. In the MCD circuit shown in Fig. 12.8, the sawtooth generator is replaced with the sawtooth generator shown in Fig. 8.1(b). (i) Draw the circuit diagrams, (ii) explain their working operations, (iii) draw waveforms at appropriate places, and (iv) deduce the expression for their output voltages.

# 13 Applications of MCDs

Multiplier-cum-dividers (MCDs) find applications in balanced modulators, amplitude modulators, frequency doublers, phase angle detectors, RMS detectors, rectifiers, inductance measurement, capacitance measurement, and automatic gain control circuits. These applications are described in this chapter.

## 13.1 BALANCED MODULATOR

Modulation and frequency doubling are basically processes of multiplication. Modulation is a process by which some characteristics of one wave called a carrier are varied in accordance with some characteristics of another wave called a modulating signal. MCDs are used for modulators and are described as below:

The MCD as a balanced modulator or suppressed carrier double side band modulator is shown in Fig. 13.1

$V_2$ is the modulating signal: $V_2 = E_m Sin\omega_m t$

$V_3$ is the carrier signal: $V_3 = E_c Sin\omega_c t$

A constant reference voltage $V_R$ is applied to the $V_1$ input of MCD. Its output will be

$$V_O = \frac{V_2 V_3}{V_1} = \frac{(E_m Sin\omega_m t)(E_c Sin\omega_c t)}{V_R} \tag{13.1}$$

$$V_O = \frac{E_m E_c}{2V_R}[\cos(\omega_c - \omega_m)t - \cos(\omega_c + \omega_m)t] \tag{13.2}$$

In the above equation the $V_R$ is usually 10. The carrier frequency term does not appear and hence the name suppressed carrier is obtained. Balanced modulators are widely used in communication systems, measurements, and instrumentation and control systems. They have the advantage over other modulation schemes in that the carrier is suppressed and does not appear in the output and hence power consumption is reduced. Modulating circuits have their spectrum centered about the second harmonic of the carrier frequency or any multiple of it and require complex filters to eliminate the unwanted frequencies.

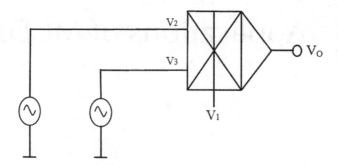

**FIGURE 13.1** MCD as a balanced modulator

A common problem in communication is to extract information from the single sideband (ss) signals received. The (ss) signal can be written as

$$e_{ss} = K \sin(\omega_m + \omega_c)t \quad (13.3)$$

If $e_{ss}$ is multiplied by an appropriate carrier signal $A \sin\omega_c t$ the resulting output will be

$$V_O = \frac{[K \sin(\omega_m + \omega_c)t](A\sin \omega_c t)}{V_R}$$

$$V_O = \frac{KA}{2V_R}[\cos \omega_m t - \cos(\omega_m + 2\omega_c)t] \quad (13.4)$$

In the above equation the term $KA/V_R \cos \omega_m t$ can be extracted by using a simple filter to remove the second high frequency term.

## 13.2 AMPLITUDE MODULATOR

Similar to a balanced modulator, when a DC voltage is added to the modulating signal, the MCD performs amplitude modulation and is shown in Fig. 13.2. In this the carrier is passed through the multiplier when the modulating signal is zero.

From Fig. 13.2

$$V_2 = E_m + mE_m \sin \omega_m t = E_m(1+m)\sin \omega_m t$$

$$V_3 = E_c \sin \omega_c t$$

# Applications of MCDs

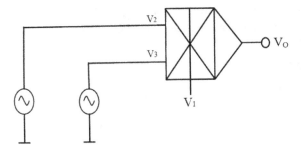

**FIGURE 13.2** MCD as an amplitude modulator

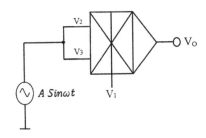

**FIGURE 13.3** Frequency doubler using MCD

Let $V_1 = V_R$

$$V_O = \frac{V_2 V_3}{V_1} = \frac{(E_m + mE_m \sin \omega_m t)(E_c \sin \omega_c t)}{V_R} \qquad (13.5)$$

$$V_O = \frac{E_m E_c}{V_R}[\sin \omega_c t + \frac{m}{2}\cos(\omega_c - \omega_m)t - \frac{m}{2}\cos(\omega_c + \omega_m)t \qquad (13.6)$$

Where m = modulation index. If we keep the peak amplitude of the modulating wave equal to the DC offset voltage a 100% modulation can be achieved.

## 13.3 FREQUENCY DOUBLER

The frequency doubler operation using a MCD is shown in Fig. 13.3
Let $V_2 = V_3 = A \sin \omega t$, $V_1 = V_R$

$$V_O = \frac{V_2 V_3}{V_1} = \frac{(A \sin \omega t)^2}{V_R} \qquad (13.7)$$

$$V_O = \frac{A^2}{2V_R}(1 - \cos 2\omega t) \qquad (13.8)$$

The multiplier output $V_O$ contains a DC voltage in association with the second harmonic of the input signal and this DC voltage can be removed through AC coupling.

## 13.4 PHASE ANGLE DETECTOR

Fig. 13.4 shows the MCD in connection with a low pass filter to determine the phase difference between two sinusoidal signals of same frequency.

The principle of the circuit is based on the trigonometric identity

$$\sin \alpha \sin \beta = \frac{1}{2}[\cos(\alpha-\beta) - \cos(\alpha+\beta)]$$

The waveform at the output of multiplier $V_X$ consists of an AC voltage superimposed on a DC level. The frequency of the AC term is twice that of input signals and the DC voltage level is proportional to the phase difference between the two inputs.

Let $V_2 = A \sin \omega t$

$$V_3 = B \sin(\omega t + \theta)$$

and $V_1 = V_R$

$$V_X = \frac{V_2 V_3}{V_R} = \frac{AB}{2V_R}[\cos\theta - \cos(2\omega t + \theta)] \qquad (13.9)$$

The RC low pass filter eliminates the AC term and allows only the DC voltage $V_O$ and is given as

$$V_O = \frac{AB}{2V_R}\cos\theta \qquad (13.10)$$

$$\theta = \cos^{-1}\frac{2V_R V_O}{AB}$$

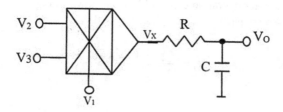

**FIGURE 13.4** Phase detector using MCD

## 13.5 RMS DETECTOR

Fig. 13.5 shows the RMS detector using an MCD.
The RMS value of a signal $V_I$ is given as

$$V_{RMS} = \sqrt{\frac{1}{T}\int_0^T (V_I)^2 \, dt} \qquad (13.11)$$

Let $V_1 = V_R$ a constant voltage. The input signal $V_I$ is squared by the multiplier-cum-divider $MCD_1$ and the squared signal $V_X$ is given to integrator $OA_1$. The integrator $OA_1$ integrates $V_X$ and gives $V_Y$ as the integrated output. The integrated output $V_Y$ is given the square rooter realized by multiplier-cum-divider $MCD_2$ and op-amp $OA_2$. The output of the square rooter is the actual RMS output $V_O$.

## 13.6 RECTIFIER

The full wave rectifier using MCD is shown in Fig. 13.6. The op-amp $OA_1$ acts as a zero crossing detector and converts the input sine wave into a square wave. Let $V_1 = V_R$. The MCD will work as multiplier.

The multiplier is a four-quadrant multiplier whose output is configured as always positive. The multiplier output is the full wave rectified output as shown in Fig. 13.7

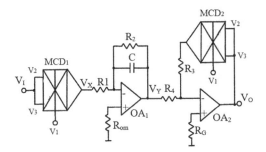

**FIGURE 13.5** Circuit diagram of RMS detector

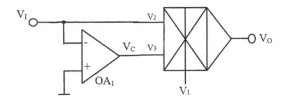

**FIGURE 13.6** MCD as full wave rectifier

## 13.7 INDUCTANCE MEASUREMENT BY PHASE ANGLE RESPONSE

The block diagram of inductance measurement by the phase angle response is shown in Fig. 13.8. The unknown inductor is connected in series with a known standard resistance $R_S$ and the combination is passed by a constant current source $I_S$ of constant frequency. The phasors of the voltages $V_S$, $V_R$, and $V_L$ are indicated in Fig. 13.9. where Vs is the source voltage, $V_R$ is the voltage across the resistor, and $V_L$ is the voltage across the unknown inductor.

It can be deduced from Fig. 13.9 that

$$L = \frac{R}{\omega} \frac{Sin\theta_1 Sin\theta_2}{Sin\theta_3} \qquad (13.12)$$

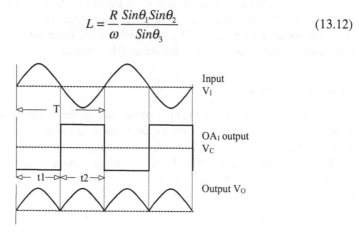

**FIGURE 13.7** Associated waveforms of Fig. 13.6

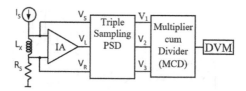

**FIGURE 13.8** L measurement phase angle response

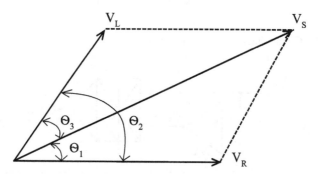

**FIGURE 13.9** Phasor diagram for inductive impedance

Applications of MCDs

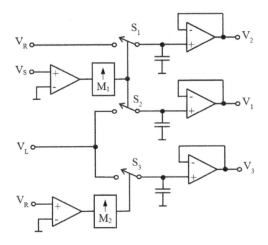

**FIGURE 13.10** Triple sampling type phase sensitive detectors

The triple sampling phase sensitive detectors is shown in Fig. 13.10. A phase shifted signal $V_R$ is sampled by a sampling pulse generated during positive zero crossings of the reference signal $V_S$. Then the sampled output will be

$$V_2 = V_R \sin \theta_1 \qquad (13.13)$$

Where $V_R$ is the peak value of the phase shifted signal and $\theta_1$ is the phase difference between $V_S$ and $V_R$. Similarly

$$V_1 = V_L \sin \theta_3 \qquad (13.14)$$

$$V_3 = V_L \sin \theta_2 \qquad (13.15)$$

In Fig.13.8 the MCD output will be

$$V_O = \frac{V_2 V_3}{V_1} = V_R \frac{\sin\theta_1 \sin\theta_2}{\sin\theta_3} \qquad (13.16)$$

On comparing equations (13.12) and (13.16), the MCD output is proportional to unknown inductance and is displayed in the digital voltmeter (DVM)

## 13.8 CAPACITANCE MEASUREMENT BY PHASE ANGLE RESPONSE

The C measurement by phase angle response is given in Fig. 13.11, the unknown capacitor is connected in series with a known standard resistance $R_S$ and the combination is passed by a constant current source $I_S$ of constant frequency. The phasors of the voltages $V_S$, $V_R$, and $V_C$ are indicated in Fig. 13.12. where Vs is the

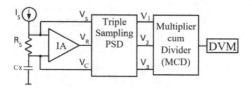

**FIGURE 13.11** C measurement phase angle response

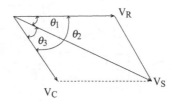

**FIGURE 13.12** Phasor diagram of Fig. 13.11

source voltage, $V_R$ is the voltage across the resistor and $V_C$ is the voltage across the unknown capacitor

It can be deduced from Fig. 13.12 that

$$C_p = \frac{1}{R\omega} \frac{\sin\theta_3 \sin\theta_2}{\sin\theta_1} \tag{13.17}$$

The triple sampling phase sensitive detectors is shown in Fig. 13.13. A phase shifted signal $V_R$ is sampled by a sampling pulse generated during positive zero crossings of the reference signal $V_S$. Then the sampled output will be

$$V_2 = V_R \sin\theta_1 \tag{13.18}$$

Where $V_R$ is the peak value of the phase shifted signal and $\theta_1$ is the phase difference between $V_S$ and $V_R$. Similarly

$$V_1 = V_C \sin\theta_3 \tag{13.19}$$

$$V_3 = V_C \sin\theta_2 \tag{13.20}$$

In Fig. 13.11, the MCD output is given as

$$V_O = \frac{V_2 V_3}{V_1} = V_R \frac{\sin\theta_1 \sin\theta_2}{\sin\theta_3} \tag{13.21}$$

On comparing equations (13.17) and (13.21), The MCD output voltage is proportional to unknown capacitance and is displayed in the digital voltmeter (DVM).

# Applications of MCDs

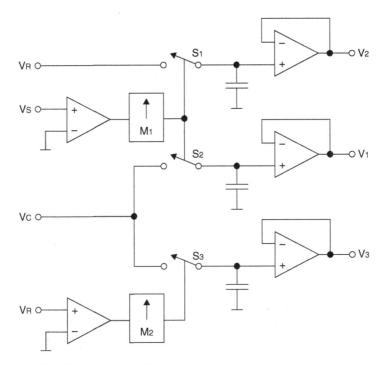

**FIGURE 13.13**  Triple sampling PSDs

## 13.9 AUTOMATIC GAIN CONTROL CIRCUIT – TYPE I

The block diagram of AGC circuit is shown in Fig. 13.14. An MCD is used as a voltage controlled linear amplifier. Let $V_1 = V_R$. The MCD works as a multiplier.

The amplitude of the output signal $V_O$ is rectified and then filtered to get a DC component only. The DC voltage at the output of LPF is compared with a reference voltage through the integrator to generate an error signal. The error signal is integrated in the high gain integrator. When the DC component of the output voltage $V_O$ is equal to the reference voltage, the integrator input is zero and the output of the integrator is steady. The output of the integrator is multiplied by the input signal $V_I$ and as varying the gain.

The amplitude stability of the output signal depends on the stability of the DC reference voltage and the integrator gain. The rate at which the loop can realize after sudden changes in input level depends on the cut off frequency of the low pass filter and the integrator time constant. AGC circuits are widely used to stabilize the signal amplitude of oscillation to keep a signal's amplitude constant while its phase angle is varied by filtering and for various other purposes.

## 13.10 AUTOMATIC GAIN CONTROL CIRCUIT – TYPE II

The MCD is used in automatic gain control system as shown in Fig. 13.15 for positive reference voltage. Let $V_3 = 10V$. The MCD works as a divider.

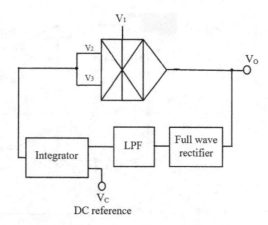

**FIGURE 13.14** AGC using MCD as multiplier for positive reference voltage

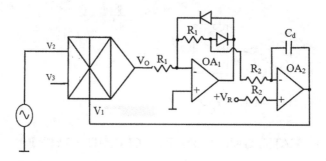

**FIGURE 13.15** AGC using MCD as divider for positive reference voltage

The numerator of the divider accepts bipolar voltages and the denominator accepts only negative voltages. The output voltage $V_O$ is half wave rectified to a DC reference by an integrator $OA_2$. The integrator output is given the denominator of divider as control signal. Therefore, variations in the output $V_O$ caused by the input voltage changes will create a control signal that corresponds for the input voltage changes. Under balanced condition

$$\frac{V_P}{\pi} = V_R \qquad (13.22)$$

Where $V_P$ = peak value of input signal and
$V_R$ = reference voltage.

# 14 Circuit Simulation

The multiplier-cum-divider (MCD) circuits described in this book can be simulated and verified using TINA software. TINA is one of the most powerful and best converging Spice simulators on the market. It includes both Berkeley Spice and XSpice-based Spice engines, and supports most Spice dialects with parallelized processing and precompiled models. In addition to the large Spice component libraries in TINA, you can create new TINA components from any Spice subcircuit, whether created by yourself, downloaded from the Internet, obtained from a manufacturer's CD or from portions of schematics turned into subcircuits. TINA automatically represents these subcircuits as a rectangular block, but you can create any shape you like with TINA's Schematic Symbol Editor. You can also use TINA's parameter extractor program to calculate model parameters from catalog or measurement data and then add the new devices into the catalog.

TINA Design Suite is a powerful yet affordable circuit simulator, circuit designer, and PCB design software package for analyzing, designing, and real-time testing of analog, digital, IBIS, HDL, MCU, and mixed electronic circuits and their PCB layouts. You can also analyze SMPS, RF, communication, and optoelectronic circuits; generate and debug MCU code using the integrated flowchart tool; and test microcontroller applications in a mixed circuit environment. Off-line licenses of TINA include free private on-line licenses for one year. You can analyze your circuit through more than 20 different analysis modes or with 10 high tech virtual instruments. Present your results in TINA's sophisticated diagram windows, on virtual instruments, or in the live interactive mode where you can even edit your circuit during operation, develop, run, debug, and test HDL and MCU applications. Electrical engineers will find TINA an easy-to-use, high-performance tool, while educators will welcome its unique features for the training environment.

With the TINACloud on-line circuit simulator, in addition to the installable versions, now you can also edit and run your schematic designs and their PCB layouts online on PCs, Macs, thin clients, tablets, smartphones, smart TVs, and e-book readers without any installation. You can use TINACloud in the office, classroom, at home and while travelling, and anywhere in the world that has Internet access.

As a software company with extensive experience in circuit simulation and as a company that licenses circuit simulator software to the largest semiconductor companies in the world, DesignSoft has a unique insight as well as experience in developing device models. Their software and modeling capability supports SPICE, VHDL, Verilog, Verilog-AMS, SystemC models, and mixed circuit models of modern integrated circuits. They create syntax-compatible models for TINA, PSpice, SIMetrix, LTSpice, and other simulators. Devices include Opamps, discrete semiconductors (diodes, MOSFETs, IGBTS, etc.), SMPS ICs, LED drivers, AD and DA converters, microcontrollers (ARM, PIC, AVR, XMC, etc.), and more. As the developer of TINA, one of the fastest simulators on the market, we have a deep understanding of the simulation algorithms allowing us to advance accurate and time efficient device models.

TINA is free to download. Please see the site below.

www.ti.com/tool/tina-ti

## 14.1 SIMULATION OF TIME-DIVISION MULTIPLY–DIVIDE MCD

The time-division multiply–divide MCD discussed in Chapter 8, Section 8.4, and shown in Fig. 8.11(b) can be simulated using TINA software. The simulated circuit diagram is shown in Fig. 14.1. The following component values are chosen. $R_1 = 10$ M, $R_3 = 10$ K, $R_4 = 100$ E, $R_5 = R_6 = 100$K, $C_1 = 39$p, $C_2 = C_3 = 100$ nF and $C_4 = 10$ uF. The observed waveforms are shown in Fig. 14.2 and the verified theoretical waveforms shown in Fig. 14.2. $V_3 = 500$ mV, $V_2 = 2.5$ V and $V_1 = 250$ mV are applied to the circuit. The theoretical output voltage is 5 V and the simulated output voltage is 5.25 V. The simulated output voltage measured by a virtual multimeter is shown in Fig. 14.3.

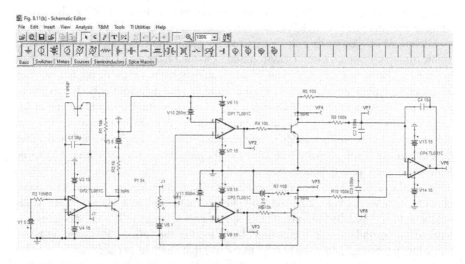

**FIGURE 14.1** Simulated design of Fig. 8.11(b)

# Circuit Simulation

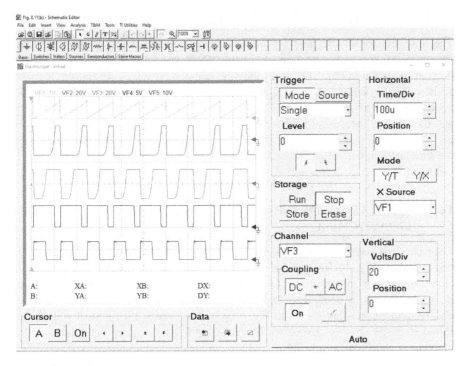

**FIGURE 14.2** Observed waveforms of Fig. 14.1

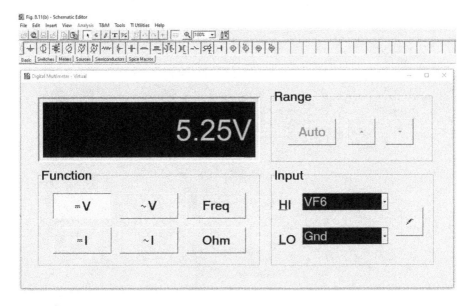

**FIGURE 14.3** Output voltage of Fig. 14.1

## 14.2 SIMULATION OF TIME-DIVISION DIVIDE–MULTIPLY MCD

The time-division divide–multiply MCD discussed in Chapter 9, Section 9.2, and shown in Fig. 9.6(a) can be simulated using TINA software. The simulated circuit diagram is shown in Fig. 14.4. The following component values are chosen. $R_1 = 33$ K, $R_2 = R_7 = R_9 = R_{10} = R_3 = 10$ K, $R_4 = 5.6$ K, $R_5 = 2.5$ K, $R_6 = R_8 = 1$ K, $C_1 = 100$ nF, $C_2 = C_3 = 1$ uF and $C_4 = 10$ uF. The observed waveforms are shown in Fig. 14.5 and the verified theoretical waveforms shown in Fig. 9.5. $V_3 = -7.5$ V, $V_1 = 7.5$ V and $V_2 = 4.5$ V are applied to the circuit. The theoretical output voltage is $-4.5$V and the simulated output voltage is $-4.54$V. The simulated output voltage measured by a virtual multimeter is shown in Fig. 14.6.

## 14.3 SIMULATION OF TIME-DIVISION MCD – TYPE I – SWITCHING

The time-division MCD – Type I – switching discussed in Chapter 11, Section 11.3, and shown in Fig. 11.7(a) can be simulated using TINA software. The simulated circuit diagram is shown in Fig. 14.7. The following component values are chosen. $R_1 = 1$ M, $R_2 = R_6 = 10$ K, $R_3 = 15$ K, $R_4 = R_5 = 1$K, $C_1 = 39$ pF and $C_2 = 1$ uF. The observed waveforms are shown in Fig. 14.8 and the verified theoretical waveforms shown in Fig. 11.6. $V_3 = 5$ V, $V_2 = 2.5$ V and $V_1 = 5$ V are applied to the circuit. The theoretical output voltage is 2.5 V and the simulated output voltage is 2.51 V. The simulated output voltage measured by a virtual multimeter is shown in Fig. 14.9.

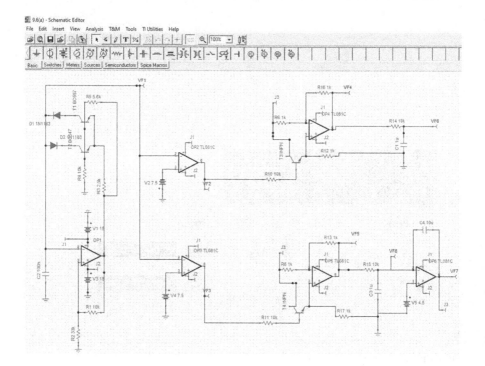

**FIGURE 14.4** Simulated diagram of Fig. 9.6(a)

# Circuit Simulation

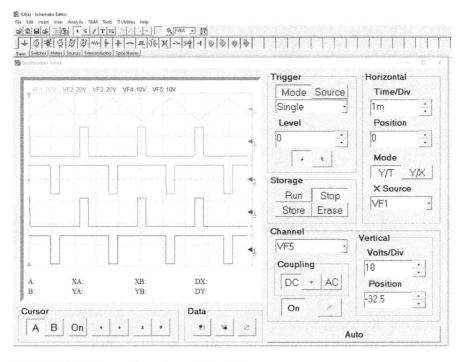

**FIGURE 14.5** Observed waveforms of Fig. 14.4

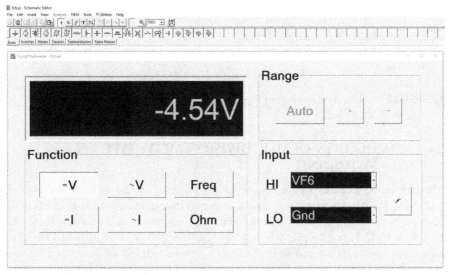

**FIGURE 14.6** Output voltage of Fig. 14.4

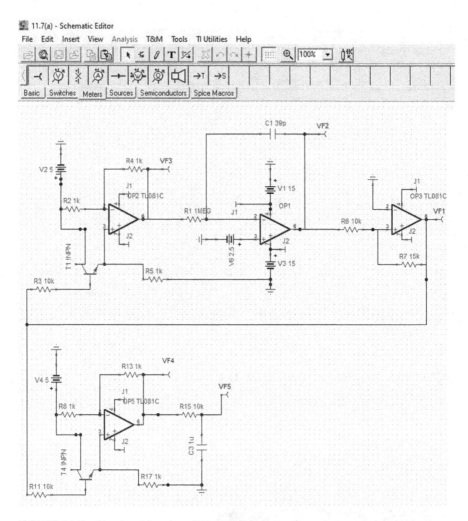

**FIGURE 14.7** Simulated drawing of Fig. 11.7(a)

## 14.4 SIMULATION OF TIME-DIVISION MCD – TYPE II – SWITCHING

The time-division MCD – Type II – switching discussed in Chapter 11, Section 11.4, and shown in Fig. 11.10(a) can be simulated using TINA software. The simulated circuit diagram is shown in Fig. 14.10. The following component values are chosen. $R_1 = 1$ M, $R_2 = R_6 = 10$ K, $R_3 = 15$ K, $R_5 = R_4 = 1$ K, $C_1 = 39$ p, $C_2 = 1$ uF and $C_3 = 100$ nF. The observed waveforms are shown in Fig. 14.11 and the verified theoretical waveforms shown in Fig. 11.9. $V_3 = 5$ V, $V_2 = 2.5$ V and $V_1 = 5$ V are applied to the circuit. The theoretical output voltage is 2.5 V and the simulated output voltage is 2.5 V. The simulated output voltage measured by a virtual multimeter is shown in Fig. 14.12.

# Circuit Simulation

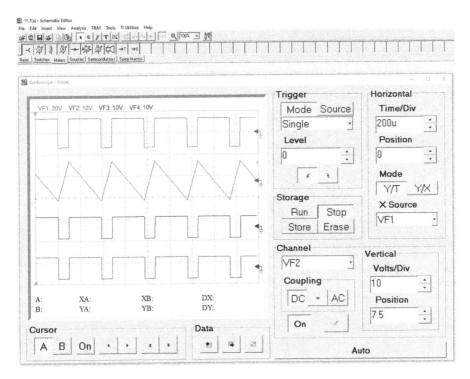

**FIGURE 14.8**  Observed waveforms of Fig. 14.7

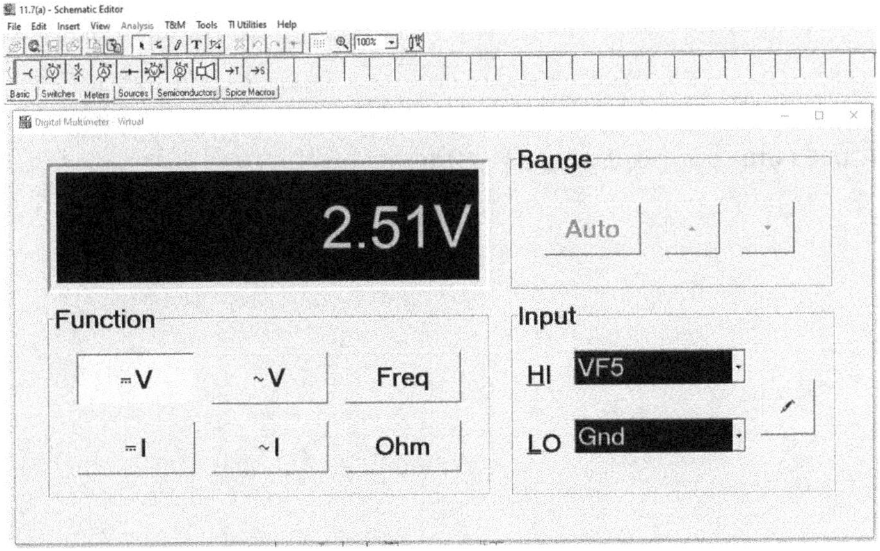

**FIGURE 14.9**  Output voltage of Fig. 14.7

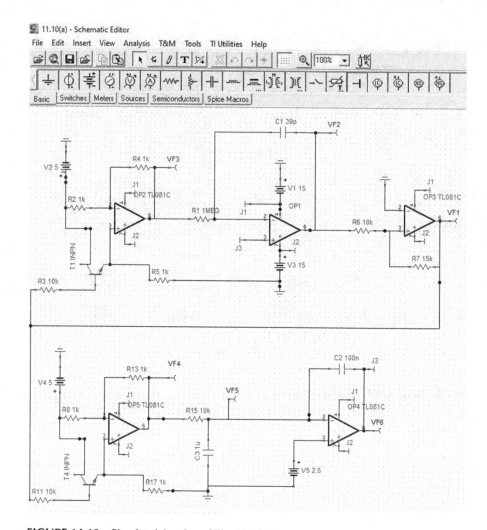

**FIGURE 14.10** Simulated drawing of Fig. 11.10(a)

# Circuit Simulation

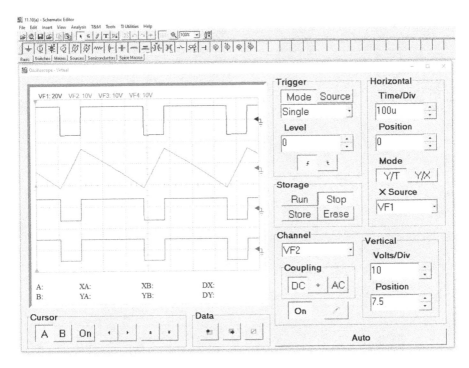

**FIGURE 14.11** Observed waveforms of Fig. 14.10

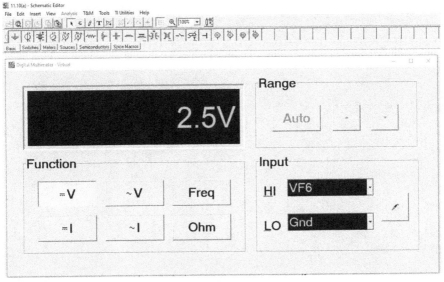

**FIGURE 14.12** Output voltage of Fig. 14.10

# Appendix

## PIN DETAILS OF ICS

The pin details of (i) operational amplifier LF 356 IC are shown in Fig. 1, (ii) analog multiplexer CD 4053 IC in Fig. 2, (iii) analog switch CD 4066 IC in Fig. 3, (iv) transistor BC547/557 in Fig. 4, and (v) CD4528 MONO in Fig. 5.

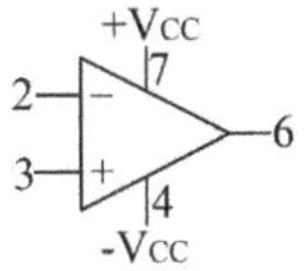

**FIG. 1** Pin details of LF 356 IC

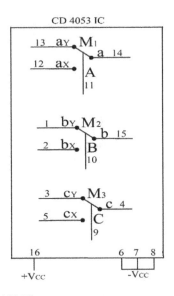

**FIG. 2** Pin details of CD 4053 IC

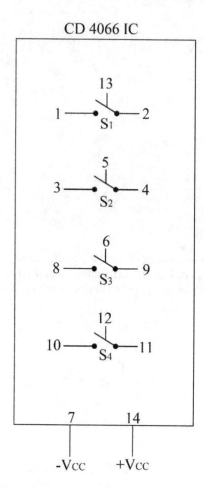

**FIG. 3** Pin details of CD 4066 IC

**FIG. 4** Pin details of BC547/BC557

# Appendix

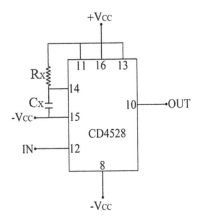

**FIG. 5(A)** Pin details of MONO CD4528 (rising edge triggering)

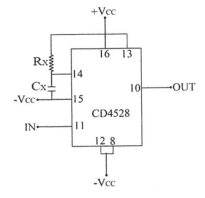

**FIG. 5(B)** Pin details of MONO CD4528 (falling edge triggering)

# Bibliography

[1] Yu jen Wong and Williams E. Ott, *Function circuits; Design and applications*, McGraw Hill, 1976.
[2] Sergio Franco, *Design with operational amplifiers and analog integrated circuits*, Tata McGraw-Hill edition, 2002.
[3] C. Selvam, "A simple and low cost pulse time multiplier," *IETE Students Journal*, Vol. 34, Nos 3 and 4, July–December 1993, pp. 207–211.
[4] C. Selvam and V. Jagadeesh Kumar, "A simple and inexpensive implementation of time division multiplier for two quadrant operation," *IETE Technical Review*, Vol. 12. No. 1, Jan–Feb 1995, pp. 33–35.
[5] K.C. Selvam, "Some techniques of analog multiplication using op-amp based sigma generator," *IETE Journal of Education*, Vol. 55, Issue 01, Jan–June 2014, pp. 33–39.
[6] K.C. Selvam, "Four quadrant time division multiplier without using any reference clock,l" *IETE Journal of Education*, November 2017, DOI: 10.1080/09747338.2017.1391719.
[7] K.C. Selvam, "A pulse position sampled multiplier," *IETE Journal of Education*, Vol. 54, Issue 1, Jan–Jun 2013, pp. 5–8.
[8] C. Selvam and V. Jagadeesh Kumar, "A simple multiplier and squarer circuit," *IETE Students Journal*, Vol. 37. Nos 1 and 2, Jan–Mar 1996, pp. 7–10.
[9] K.C. Selvam, "A double single slope multiplier cum divider," *IETE Journal of Education*, Vol 41, Nos 1 and 2, Jan–June 2000, pp. 3–5.
[10] K.C. Selvam, "Double dual slope multiplier-cum-divider," *Electronics Letters*, Vol. 49, No. 23, 7 November 2013, pp. 1435–1436.
[11] M. Tomata, Y. Sugiyamma, and K. Yamaguchi, "An electronic multiplier for accurate power measurement," *IEEE Trans.Instrum.Meas*, Vol IM-17, 1968, pp. 245–251.
[12] T.S Rathore and B.B Bhatacharya, "A Novel type of Analog Multiplier," *IEEE Trans on Industrial Electronics*, Vol IE-31, 1984, pp 268–271.
[13] Greg Johnson, "Analysis of modified Tomata-Sugiyama-Yamaguchi multiplier," *IEEE Transactions on Instrumentation and Measurement*, Vol. 33, No, 1, 1984, pp.11–16.
[14] G. Han and E. Sanchez-Sinencio, "CMOS transconductance multipliers; A tutorial," *IEEE CAS II; Analog and digital signal processing*, Vol.45 no 12, Dec. 1998, pp. 1550–1563.
[15] N. Bajaj and Jivesh Govil, "Realisation of a differential multiplier-divider based on current feedback amplifiers," *Semiconductor Electronics*, 2006. ICSE '06. IEEE International Conference on DOI:10.1109/SMELEC.2006.380728, pp. 708 – 712.
[16] Weixin Gai, Hongyi Chen, and E. Seevinck, "Quadratic-translinear CMOS multiplier-divider circuit," *Electronics Letters*, Vol 33, No. 10, May 1997, pp. 860–861

[17] M.S. Piedade, "New analogue *multiplier-divider* circuit based on cyclic data convertors," *Electronics Letters*, Vol 26, No. 1, January 1990, pp. 2–4.
[18] A.J. Peyton and V. Walsh, *Analog electronics with op-amps*, Cambridge University Press, 1993.
[19] George Clayton and Steve winder, *Operational amplifiers*, Fifth edition, Elsevier publication, 2003, pp. 306–307.
[20] D. Roy Choudhury and Shail B. Jain, *Linear integrated circuits*, Third edition, New Age International Publishers, 2010, pp. 220–222.

## FUNCTION CIRCUITS TUTOR KIT

The author designed and developed the "Function Circuits Tutor Kit" to practically verify his concepts and theory on function circuits. The kit has all the components of function circuits as individual blocks. The respective blocks can be connected by external wires or patch cards to form a required function circuit. The kit has internally generated input voltages $V_1$, $V_2$ and $V_3$. Multimeters, power supplies and an oscilloscope are required externally to experiment the function circuit. On reading this book, the reader will be able to verify the multiplier-cum-divider circuits by experimenting with this trainer kit. This experimental kit is a very useful full kit that each and every electronics laboratory in the world should have. The photograph of this kit is given in the next page.

For further details and to get this kit contact: kcselvam@ee.iitm.ac.in

# Bibliography

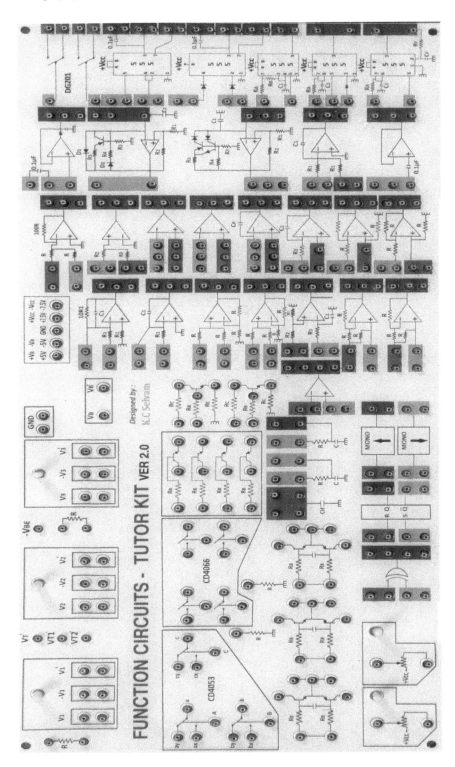

# Index

Amplitude modulator 188
Analog multiplexer 14
Analog switches 11
Anti log amplifier 32
Astable multivibrator 15
Automatic gain control – type I 195
Automatic gain control – type II 195

Balanced modulator 187

Capacitance measurement 193

Differential integrator 11
Divide – multiply MCD 119
Double dual slope MCD with FBC 133
Double dual slope MCD with FBC 81
Double dual slope MCD with Flip-Flop 140
Double dual slope MCD with Flip-Flop 85
Double multiplexing – averaging MCD 53
Double Single slope MCD 129
Double Single slope MCD 77
Double switching – averaging MCD 97

Flip flop 85
Frequency doubler 189
Full wave rectifier 26

Half wave rectifier 25

Inductance measurement 192
Integrator 10
Inverting amplifier 7
Inverting Schmitt trigger 24

Log amplifier 30
Log-antilog MCD – type I 37
Log-antilog MCD – type II 40

MCD using FETs 41
MCD using MOSFETs 43
MCD using voltage tunable astable multivibrator 91
MCDs using multipliers and dividers 46
Multiply – divide MCD 123

Non-inverting amplifier 8
Non-inverting Schmitt trigger 24

Peak detecting MCD using voltage tunable astable multivibrator 149
Peak detector 29
Phase angle detector 190
Practical integrator 10
Pulse position peak detecting MCD – multiplexing 173
Pulse position peak detecting MCD – switching 177
Pulse position peak sampling MCD – multiplexing 180
Pulse position peak sampling MCD – switching 183
Pulse width integrated MCD 145
Pulse width integrated MCD 88

Rectifier 191
RMS detector 191

Sample and hold circuit 30
Saw tooth generators using analog switches 95
Saw tooth wave generators using multiplexers 51
Schmitt trigger 24

Time division divide – multiply MCD 109
Time division divide – multiply MCD 61
Time division divide – multiply MCD 70
Time division MCD 115
Time division MCD 68
Time division MCD with no reference – type I – Multiplexing 155
Time division MCD with no reference – type I – Switching 161
Time division MCD with no reference – type II – Multiplexing 158
Time division MCD with no reference – type II – Switching 166
Time division multiply – divide MCD 104
Time division multiply – divide MCD 58
Time division multiply – divide MCD 73
Time division single slope peak detecting MCD 101
Time division single slope peak detecting MCD 55
Triangular wave generators 65

Voltage comparator 19
Voltage to frequency converter 16
Voltage to period converter 16

Window detector 34

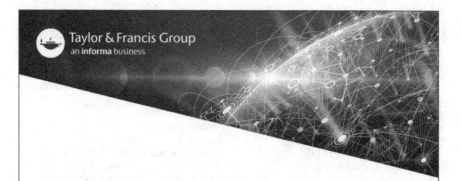